*Origins of
the Science
of Crystals*

JOHN G. BURKE

ORIGINS
of
THE SCIENCE
of CRYSTALS

UNIVERSITY OF CALIFORNIA PRESS

BERKELEY AND LOS ANGELES

1966

University of California Press
Berkeley and Los Angeles, California
Cambridge University Press
London, England
© 1966 by The Regents of the University of California
Library of Congress Catalog Card Number: 66-13584
Designed by John Goetz
Printed in the United States of America

Acknowledgments

I am sincerely grateful for the comments and suggestions offered by Professors Adolph Pabst, University of California, Berkeley; Cecil Schneer, University of New Hampshire; Cyril Stanley Smith, Massachusetts Institute of Technology; and Daniel F. Weill, University of California, San Diego, who read the manuscript of this study at various stages during its preparation. I am also in debt to Professor Gary Ernst, University of California, Los Angeles, whom I consulted concerning certain technical points. Any errors, however, are the sole responsibility of the author.

Also, I wish to thank G. Bell & Sons, Ltd.; Dover Publications, Inc.; McGraw-Hill Book Company; the Royal Dublin Society; the Swedenborg Society; and John Wiley & Sons, Inc., for their permission to use illustrative material.

Finally, I wish to thank my wife, Peggy, not only for her comments, but also for her unfailing patience and understanding.

Contents

Abbreviations

ADB	*Abhandlugen der königlichen Akademie der Wissenschaften in Berlin*
ADC	*Annales de Chimie et de Physique*
ADS	*Mémoires de l'Académie Royale des Sciences*
AMHN	*Annales du Museum d'Histoire Naturelle*
AP	*Annals of Philosophy*
EPJ	*Edinburgh Philosophical Journal*
GJ	*Journal für die Chemie und Physik* von A. F. Gehlen
Haüy, *Essai*	R. J. Haüy, *Essai d'une théorie sur la structure des crystaux* (Paris, 1784)
Haüy, *TDC*	R. J. Haüy, *Traité de cristallographie* (3 vols.; Paris, 1822)
Haüy, *TDM1*	R. J. Haüy, *Traité de minéralogie* (5 vols.; Paris, 1801)
Haüy, *TDM2*	R. J. Haüy, *Traité de minéralogie* (2d ed.; 5 vols.; Paris, 1822)
JDP	*Journal de Physique*
JHN	*Journal d'Histoire Naturelle*
PM	*Philosophical Magazine*
PT	*Philosophical Transactions of the Royal Society of London*
SJ	*Journal für Chemie und Physik* von J. S. C. Schweigger

CHAPTER I

Introduction

The modern development of the science of crystals began in 1912 when, at the suggestion of Max von Laue (1879-1960), professor of physics at the University of Munich, a research assistant, Walter Friedrich, and a doctoral candidate, Paul Knipping, passed a narrow beam of X rays through a crystal of copper sulfate and recorded a pattern of spots on a photographic plate (fig. 1). Shortly thereafter, from a study of another pattern produced by passing the beam through zinc blende, von Laue was able to demonstrate that it could be due only to the diffraction of very short waves by a regular arrangement of atoms or molecules in the crystal, a conclusion that confirmed the recognized cubic symmetry of the substance.[1] Within a short time, William H. Bragg (1862-1942) and his son, William Lawrence Bragg (1890——), utilized and extended this diffraction method to determine the arrangement of the atoms within such simple crystalline materials as common salt, pyrite, fluorite, and calcite.[2] Since that time, the improvement of the techniques of X-ray crystallography has resulted in an enor-

[1] W. Friedrich, P. Knipping, and M. von Laue, "Interferenzen-Erscheinungen bei Röntgenstrahlen," *Sitzungsberichte der mathematisch-physikalischen Klasse der K. B. Akademie der Wissenschaften zu München* (1912), pp. 303-322. P. P. Ewald, ed., *Fifty Years of X-ray Diffraction* (Utrecht, 1962), pp. 31-80, gives a complete account of von Laue's discovery and the immediate sequels to it. Also of interest is M. von Laue, "Historical Introduction," in *International Tables for X-ray Crystallography* (Birmingham, 1952), I, 1-5.

[2] W. L. Bragg, "The Diffraction of Short Electromagnetic Waves by a Crystal," *Proceedings of the Cambridge Philosophical Society*, XVII (1913), 43. Other early articles by the Braggs appeared in *Proceedings of the Royal Society of London*, Ser. A, LXXXIX (1913).

Fig. 1. The first photograph of X-ray diffraction by a crystal. SOURCE: *W. Friedrich, P. Knipping, and M. von Laue, "Interferenzen-Erscheinungen bei Röntgenstrahlen,"* Sitzungsberichte der mathematisch-physikalischen Klasse der K. B. Akademie der Wissenschaften zu München *(1912),* Heft II, fig. 1.

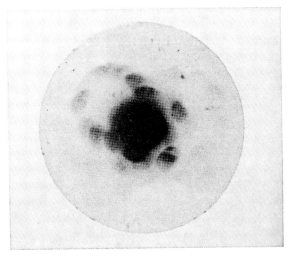

mous increase in the store of scientific knowledge of matter in the solid state, with consequent impact on the development of the sciences of physics, chemistry, biology, and geology. Crystallographers have been able to determine precisely the manner in which atoms and ions are positioned in large numbers of inorganic crystals and have extended their research to encompass the more complex molecules of organic substances. They can calculate the distances between atoms and thus infer their sizes, in this way providing information on the basis of which the type and intensity of the interatomic forces may be determined. Properties of matter which could not be explained previously on any strictly scientific basis not only can be predicted by an analysis of the crystal structure but also can be modified by the substitution of some atoms or by the addition of others. Improved techniques of crystal analysis give promise of the eventual determination of the structures of the very complex organic molecules involved in vital processes. The work of von Laue and the Braggs, then, introduced an element of certainty into the science of crystallography which had previously been lacking. By placing this powerful analytic tool in the hands of crystallographers, the determination of the internal geometry of crystals and the geometrically dependent properties was made possible, resulting in the birth of solid-state physics. It should be emphasized, however, that scientists were well prepared to utilize this new discovery. The main conceptual framework within which they worked, particularly the idea of the space lattice and the theory of space groups, had already been erected by the theoretical considerations of their predecessors.

In broad terms, the analysis by X-ray diffraction verified the contemporary theory of the structure of crystalline matter. Crystals were pictured as solids which possessed homogeneity in the sense that they were characterized by the continuous three-dimensional order of their constituent particles. Two-dimensional periodicity is expressed, for example, by patterned wallpaper on which a basic motif is repeated in two directions. If a particular point on each pattern is singled out, one can visualize a two-dimensional array of points. Similarly, if the motif is repeated in three directions, a network of points or a lattice is the result. Crystals have this kind of internal order. Atoms, ions, ionic groups, or molecules are arranged with a systematic spatial relationship to the points or nodes of a three-dimensional lattice. This internal order is reflected externally in that the faces of a crystal display certain types of symmetry. Some crystals, for example, possess perfect cubic symmetry. One can conceive of three mutually perpendicular axes of equal length passing through a cube and intersecting at its center. If the cube is rotated 90 degrees about any one of these axes, it will be brought into coincidence with its original aspect. The cube is thus designated as having three axes of fourfold symmetry. Similarly, four axes of three-fold symmetry coincide with the body diagonals of a cube, and six axes of two-fold symmetry bisect two opposite edges. The only kinds of axes of symmetry which a crystal can possess are two-fold, three-fold, four-fold, and six-fold. Crystals cannot have axes of five-fold symmetry, as do some flowers, nor axes of seven-fold symmetry, because the units that compose them must fill space. Five-sided blocks, for example, whose ends are regular pentagons, when fitted together would leave voids between them. Such units do not meet the requirement that space must be filled. In addition to symmetry about an axis, a crystal may possess the symmetry of reflection across a plane wherein one half of the crystal is a mirror image of the other half. The cube has nine planes of symmetry. Each of three is parallel to one of the pairs of opposite faces and lies midway between them. Each of the remaining six contains a pair of opposite edges. Further, a crystal may possess the symmetry of inversion through a center, if a point on its surface has a corresponding point on the opposite side equidistant from the center along an imaginary line drawn through the center and the two points. Again, the cube possesses such symmetry. Most crystals, however, are less symmetrical in a crystallographic sense, because their axes may not be mutually perpendicular, or because the axes may be of unequal length, or both. As a consequence, such crystals do not possess the combination of symmetry elements described above for the perfect cube.

In a mathematical analysis of these symmetry elements, J. F. C. Hessel (1796-1872), in 1830, found that there could be only thirty-two unique combinations.[3] These are called the symmetry classes, and the external symmetry of any crystal must correspond to one of the thirty-two. Two decades later, Auguste Bravais (1811-1863) considered the types of geometric figures formed by points distributed regularly in space and proved that points or particles could be arranged in a maximum of fourteen types of space lattices (fig. 2).[4] These lattices differ by symmetry and geometry, but in each lattice every particle is transitionally equivalent. Each particle in a particular lattice has the same environment. Further, Bravais related the space lattices to the external symmetry of crystals, that is, to the symmetry classes. Late in the century, E. S. Federov (1853-1919) and, independently, Artur Schoenflies (1853-1928) demonstrated that there were only 230 possible space groups. This is the number of symmetrical ways of arranging points in space so that the environment around each point is precisely the same as around any other point, but not necessarily

---

[3] J. F. C. Hessel, *Krystallometrie oder Krystallonomie und Krystallographie*, Ostwald's Klassiker der exakten Wissenchaften, nos. 88, 89 (Leipzig, 1897). Hessel's article was originally published in Gehler's *Physikalischen Wörterbuche* (Leipzig, 1830), *sub* "Krystall."

[4] A. Bravais, "Les systèmes formés par des points distribués regulièrement sur un plan ou dans l'espace," *Journal de l'Ecole Polytechnique*, XIX (1850), 1-128 (presented to the Academy of Sciences, Dec. 11, 1848). Amos J. Shaler, *On the Systems Formed by Points Regularly Distributed on a Plane or in Space* (Crystallographic Society of America, 1949), is an English translation of this work. Also see A. Bravais, "Etudes cristallographiques," *Journal de l'Ecole Polytechnique*, XX (1851), 101-276 (presented to the Academy of Sciences, Feb. 26, Aug. 6, 1849).

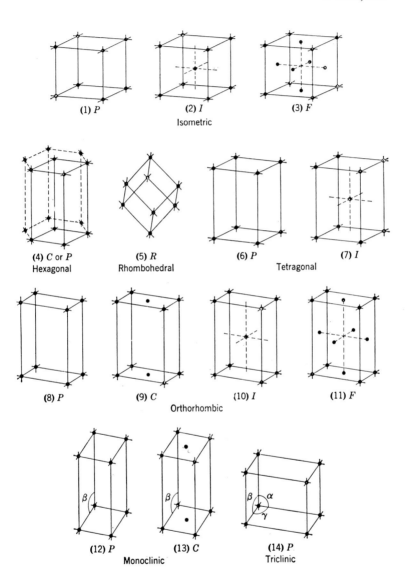

Fig. 2. Bravais space lattices. SOURCE: Dana's Manual of Mineralogy (17th ed., © 1959), p. 8. By permission of the publisher, John Wiley & Sons, Inc.

(1) P

(2) I

Isometric

(3) F

(4) C or P
Hexagonal

(5) R
Rhombohedral

(6) P

Tetragonal

(7) I

(8) P

(9) C

Orthorhombic

(10) I

(11) F

(12) P

Monoclinic

(13) C

(14) P
Triclinic

similarly oriented as in the Bravais space lattices.[5] They arrived at this result by complete consideration of combined symmetry operations, using the screw axis in which rotation about an axis is combined with translation along the axis, and also using the glide plane in which reflection in a mirror is combined with a similar translation without rotation along the axis. X-ray diffraction analysis confirmed that the atoms, ions, ionic groups, and molecules of solid matter, most of which is crystalline, were indeed arranged in the predicted ways (fig. 3).

Fig. 3. Laue photograph of beryl. SOURCE: Dana's Manual of Mineralogy *(17th ed., © 1959), p. 134. By permission of the publisher, John Wiley & Sons, Inc.*

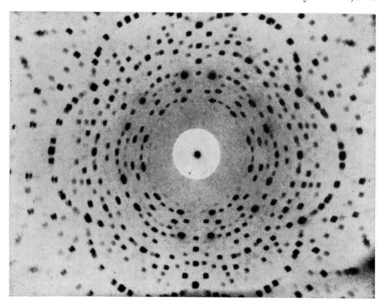

Thus, much of the work of crystallographers in the last half of the nineteenth century involved mathematical analysis. But, in turn, their results rested on earlier foundations, the discovery that the structure of crystals could be investigated mathematically in the first place, the realization that crystals evidenced symmetry,

---

[5] E. S. Federov, "Simmetriia Pravil'nykh Sistem Figur," *Zap. Min. Obshch.* ["The Symmetry of Real Systems of Configurations," *Transactions of the Mineralogical Society*], XXVIII (1891), 1-146; A. Schoenflies, *Kristallsysteme und Kristallstruktur* (Leipzig, 1891).

and the determination that all crystalline matter might be subsumed under one or another of six crystal systems.

The present study is a history of the evolution of these earlier foundations during the late eighteenth and early nineteenth centuries. Within two decades after von Laue's discovery, two important histories of the science of crystallography appeared: Hélène Metzger's *La Genèse de la science des cristaux* (Paris, 1918) and Paul Groth's *Entwicklungsgeschichte der mineralogischen Wissenschaften* (Berlin, 1926). Both are excellent and definitive studies, and both have been of great value to me not only as bibliographical sources but also in directing my attention to important aspects in the rise of crystallography to the status of a science. The first is concerned primarily with the seventeenth- and eighteenth-century speculations concerning crystalline matter which culminated in the emergence of the science of crystals in the work of René Just Haüy (1743-1822). The second describes the positive contributions to crystallography of the host of scientists from the seventeenth through the early years of the twentieth century who were concerned with crystals. It is my hope that this book will supplement these two works by bringing to light information, either not contained or little noted in them, which bears directly on the early development of the science.

But, also, there is a different emphasis. These earlier studies stressed the development of the concept of the geometry of matter and its mathematical analysis, which a few years previously had been validated. Paul Groth, in fact, was teaching crystallography at Munich and was consulted by von Laue at the time of the diffraction experiment. In the past forty years, however, the entire fields of solid-state physics and crystal chemistry have developed, wherein such problems as the binding forces within crystals, the growth of crystals, the electronic properties of crystals, and the detection, control, and modulation of light by crystals have been subjected to exhaustive study. All this work has focused attention on crystalline properties that are dependent upon the structure, but not upon the geometry, of the form as such. Most of the problems that are now being solved were faced in the early years of the science, during the lifetime of Haüy, and the questions raised by their investigation had a profound effect upon the development of the science. Studies of the anisotropic optical and electrical properties of crystals, for example, demonstrated that chemistry and geometry were not the only methods that one might use in approaching the problem of the constitution of crystalline matter. I have attempted to treat such investigations in some detail and to emphasize their relevancy to the growth of the science.

Because this study concentrates on the internal development of the science of crystals, it is my hope also that it will make a contribution in the direction of clarifying some issues raised in recent years by scholars interested in the general problem of the growth of science. I believe I have placed sufficient emphasis upon such factors as the use of the microscope in the investigations of crystals, the increasing necessity for more suitable methods of mineral identification, and the introduction of the reflecting goniometer to demonstrate that hypotheses concerning the structure of crystalline matter and its systematic classification were not unaffected by external developments, technological and otherwise. In like manner, I trust that it is apparent that other sciences, botany, biology, and geology, in addition to chemistry and optics, exerted influence on the budding science of crystals. But I have subordinated extended discussions of the effect of these features to my principal purpose of delineating the underlying assumptions, the empirical evidence, and the methodological approaches that shaped the science of crystallography at its inception and, in time, provoked important modifications in its direction.

I believe this study gives strong support to what I consider to be the central thesis of Professor Thomas Kuhn, which was presented in his recent work *The Structure of Scientific Revolutions* (Chicago, 1962). Therein Professor Kuhn maintained that the development of science, or of a science, does not occur by the gradual accumulation of scientific facts which are subjected to close scrutiny by scientists and then placed in their proper niches and later, perhaps, discarded when more facts become available. Rather, Kuhn views the progress of science as being marked by "normal" periods that occur after the achievement of a successful synthesis, when a paradigm is set forth. During the "normal" period, the majority of scientists working in the particular area are engaged in resolving problems with the purpose of extending the scope of the synthesis and proving the essential worth of the paradigm. But then, Kuhn argues, puncturing this "normal" period, there occurs an upheaval or a revolution occasioned by a reinterpretation of the available evidence or by the discovery of new evidence which causes the former synthesis to be discarded and provides the basis for a new paradigm. The early development of the science of crystals is a striking example of such a process. In the protoscientific period, from amid the maze of speculations concerning the structure of crystalline matter, a few hypotheses emerged and survived the tests of reasonableness and explanatory power. Simultaneously, there were attempts to develop rational systems of crystal classification. Haüy's achievement in bringing together evidence and hypothesis in a unified theory, wherein the variations in form of crystals of the same substance could be explained mathematically, inaugurated the initial normal period in the science of crystals. His students and adherents were primarily involved in extending his method to encompass all known crystalline substances. The science of crystals was the science of Haüy. But when the deficiencies in Haüy's theory became apparent, when scientists produced new facts contradicting certain important foundations of the science as it had been erected by Haüy, there was a fresh approach which incorporated only the undeniably valid portions of the previous one. A new paradigm was established, and thereafter crystallographers sought to extend its usefulness. In illustrating this process, I have, admittedly, stressed the weak points of Haüy's synthesis and pointed out his reluctance and, to a certain extent, his refusal to alter certain premises. But it is neither my wish nor my intention to denigrate his achievements. He was a scientist who devoted his life to the pursuit of knowledge of the physical world, who attempted and succeeded in placing the study of crystalline matter on firm scientific foundations, and, for this accomplishment, Haüy's fame in the annals of science remains secure.

CHAPTER II

Early Theories of Crystal Structure

The diversity of opinions concerning the nature of matter in the seventeenth and eighteenth centuries is perhaps nowhere more clearly illustrated than in the literature devoted to the explanation of the structure and properties of mineral and artificially produced crystals. In these writings, one can detect ideas that persisted from antiquity, for example, the view that matter was composed of a mixture of four elementary principles, fire, air, earth, and water, with qualitative differences being due to variations in the quantities of these elements. Pervading such theories were the alchemical speculations of centuries, beliefs in astral influences, animism, and superstition. But they were also beginning to be tempered by the facts of experience and by observations of chemical and physical properties of crystals which the new scientific orientation demanded be taken into consideration. Explanations of the constitution of crystals ranged from the strictly mechanistic idea that they were formed by the orderly juxtaposition of the smallest particles of inert matter to the assertion that they had life, that they were generated and grew in the same manner as vegetables and animals.

In the latter half of the eighteenth century, however, one can observe the rubble of vague speculation being cleared and notice the emergence of two distinct theories of the nature of matter. To the first I shall give the generic name "molecular," and the second I shall term "polar." While these designations are not inventions, they are unusual. I wish to employ them, first of all, in order to differentiate these theories from the atomic and field theories of the nineteenth century, which, although encompassing certain aspects of the molecular and polar concepts respectively, were nevertheless distinct. Further, I wish to avoid confusion of the ideas that concerned submicroscopic matter with the so-called mechanical and dynamical philosophies of matter. The latter were to a much greater extent involved with the nature of the force existent between gravitating masses, whether the idea of action at a distance could be maintained or whether some form of contact action was a necessary postulate. Also, it is important that a distinction be made between the two theories, because they were "regulative beliefs." The questions asked and the hypotheses suggested by investigators depended upon whether they held one theory or the other.

The molecular concept consisted essentially in the presupposition of the existence of primitive particles or atoms which were solid, impenetrable, and indestructible. The adherents of this theory believed that the atoms grouped together to form second-order particles called corpuscles or molecules of various sizes and shapes. By some, both the atoms and molecules were considered to possess innate forces of attraction and repulsion, by virtue of which they combined in certain ways. By others, the aggregations were thought to be effected by purely mechanical means, by the pressure of a subtle ether, or by the general laws of motion. But, in any event, the diverse types of combinations served to explain the qualitative differences of bodies, that is, the chemical and physical properties of the macroscopic substances to which their union gave rise. Although some supporters of the molecular theory believed that the molecules were spherical, the majority conceived of them as minute polyhedra, variously shaped but not spherical, for reasons that we shall explore.

The partisans of the polar concept, on the other hand, denied the existence of ultimate atoms, believing instead that matter was divisible to infinity. They gave first priority to the existence of forces, both attractive and repulsive, which were centered on but not simply concentrated at mathematical points. The forces emanating from these points or poles exerted influence on proximate poles giving rise to a dynamic equilibrium of forces and providing the basis for matter. Thus matter was conceived of as being constructed from these force centers, so that atoms or molecules were merely secondary, tertiary, or even more complex aggregations of the poles.

Subsumed under both theories were diverse concepts concerning the mode of aggregation of the particles and poles, the nature of the particles that resulted from the union, and other differences, which is why the rubric "generic" has been employed. But inasmuch as both concepts contributed to the development of the science of crystals, it is imperative to locate the sources of these ideas, to determine the encrustations that were added through the centuries, and to ascertain why these theories gained ascendency and in what manner they were applied to explain the constitution and properties of crystals. It is to these questions that the present chapter is addressed.

We know that Peking man collected rock crystals—transparent quartz—most probably for use as tools, but it may be that even at this early period crystals were involved in primitive animism. We do know that Australian aborigines use rock crystals and amethysts as rain stones in rain-making rites and that they also attribute to crystals sympathetic powers of a malevolent nature. There is a wealth of material to demonstrate that the belief in the medicinal and curative powers of gems and crystals was deeply rooted in Western civilization also; indeed, it still persists in the twentieth century.[1] But, although the colors, the brilliance, and the regular geometric forms of mineral crystals generated a belief in their possession of magical properties, their intrinsic beauty aroused aesthetic appreciation as well; they became gem stones which were desirable and valuable. For example, V. Gordon Childe inferred from the presence in middens of lapis lazuli, the only large deposit of which was at Badakshan in northern Afghanistan, that an active trade existed between Mesopotamia and Egypt in the predynastic period and between Mesopotamia and the Indian civilizations of the third millenium B.C.[2]

It is not until the Hellenic period, however, that records of attempts at rational explanations of the nature of matter in general

---

[1] See particularly Frank Dawson Adams, *The Birth and Development of the Geological Sciences* (New York: Dover, 1954), pp. 137-169; also Lynn Thorndike, *A History of Magic and Experimental Science* (8 vols.; New York, 1923-1958), I, 294; VI, 298-354.

[2] V. Gordon Childe, *New Light on the Most Ancient East* (New York, 1957), pp. 65, 177.

and of crystals in particular appear. These resulted from endeavors of the Greek thinkers to answer a central problem, the presence of change in the world of nature. Four theories are most germane to our subject: that of the atomists, the most famous of whom were Leucippos (*ca.* 475 B.C.), Democritus (*ca.* 470-400 B.C.), Epicurus (342-270 B.C.), and Lucretius (*ca.* 95-55 B.C.); that of Plato (427-347 B.C.); that of Aristotle (384-322 B.C.); and that of the Stoics, which is associated primarily with Zeno of Cition (*ca.* 332-262 B.C.), Chrysippus (*ca.* 280-207 B.C.), and Poseidonius (*ca.* 135-51 B.C.).

A main tenet of the atomic theory of the ancients was the existence of an infinite number of ultimately indivisible particles or atoms. As adherents of a monistic view of nature, the atomists asserted that atoms were the only reality, from which the conclusion followed that they were qualitatively identical or homogeneous. They did, however, differ from one another in size and shape, and as they moved through an endless void of nonbeing, their various juxtapositions and shifting arrangements in time accounted for the appearance, alteration, and corruption of all natural bodies. These elements entered into the arrangement, or what Robert Boyle (1627-1691), in framing the corpuscular philosophy in the seventeenth century, was to call the "texture." These were the shape, order, and position of the individual atoms of an aggregate, and thus the texture of a substance was what differentiated it qualitatively from any other substance. The shape, order, and position could be likened to the letters of the alphabet and the words and sentences formed from them. As Lucretius wrote:

> For just as all things of creation are,
> In their whole nature, each to each unlike,
> So must their atoms be in shape unlike—
> Not since few only are fashioned of like form,
> But since they all, as general rule, are not
> The same as all. Nay, here in these our verses,
> Elements many, common to many words,
> Thou seest, though yet 'tis needful to confess
> The words and verses differ, each from each.[3]

Cohesion of the atoms and aggregates was effected by mechanical means. Hooked atoms grappled those with eyes. The most com-

---

[3] William E. Leonard, trans., *Lucretius: On the Nature of Things* (New York, 1950), p. 71.

pacted bodies, the hardest, Lucretius explained, such as diamonds, must be composed of branchlike atoms which intertwined with one another.[4]

Plato's theory of matter was syncretic. From the fifth-century B.C. Pythagoreans he borrowed the notion that nature was reducible to mathematical elements, but, whereas it appears that the Pythagoreans postulated the creation of lines from mathematical points, then planes from lines, and solids from planes, Plato selected isosceles and equilateral plane triangles as the shapes from which matter was constructed.[5] Four basic solids—four of the five regular solids of classical geometry: tetrahedron, cube, octahedron, and icosahedron—were formed from these plane triangles. Within these four hollow solids, Plato imagined, existed the four powers or qualities that Empedocles (ca. 500-430 B.C.) had postulated as the basic elements of the universe. Fire was within the tetrahedron, earth within the cube, air within the octahedron, and water within the icosahedron. Plato explained that no single particles were visible because of their minuteness, but that each was capable of being separated into its composing triangles by the action of one or more of the others. The triangles might regroup into the same element or into another, or by joining other loose triangles they could become the basis of a mixed figure. There were many mixed figures, and further variety was possible because different sizes of isosceles and equilateral triangles had been used. But each mixed figure could be referred to as a member of one of four species, the perfect form of which was one of the four elements. Plato's hypothesis, then, drew attention to the specific configuration of the ultimate particles of matter. It provided a basis, albeit obscure, on which some of the chemical or physical properties of matter might be explained by reference to the shape of the particles, and it delineated how substances could exist whose properties were akin to those of the postulated four basic elements. In designating the form or shape of the ultimate particles as a primary distinguishing characteristic, Plato's theory paralleled that of the atomists. The difference lies in that for the atomists the ultimate matter was homogeneous.

In only one passage in the *Timaeus* does Plato refer to crystals:

---

[4] *Ibid.*, p. 61.
[5] B. Jowett, trans., *The Dialogues of Plato* (2 vols.; New York, 1937), II, 33-41.

Of the varieties of earth, that which has been strained through water becomes a stony substance in the following way. When the water mixed with it is broken up in the mixing, it changes into the form of air; and when it has become air, it rushes up toward its own region. But there was no empty space surrounding it, accordingly it gives a thrust to the neighboring air. This air, being heavy, when it is thrust and poured round the mass of earth, squeezes it hard and thrusts it together into the places from which the new-made air has been rising. Earth thrust together by air so as not to be soluble by water forms stone, the finer being the transparent kind consisting of equal and homogeneous particles, the baser of the opposite sort.[6]

From this account we must assume that Plato considered the most perfect transparent crystals to be aggregates of the cubic particles of earth. But his poetic imagination precluded any clear account as to why substances cohered when they were removed from the surrounding earth.

The reasons for Aristotle's rejection of Plato's theory were clear and concise, based primarily on his conviction that a void could not exist in nature.[7] There were only two solids that could fill space, he asserted, the tetrahedron and the cube, but the theory needed more than these because the elements it recognized were more in number. Also, it was clear, Aristotle asserted, that simple bodies are often given a shape by the place in which they were formed, and, in these instances, the shape of the element could not persist, for otherwise the contained mass would not be in continuous contact with the containing body. Further, there were continuous bodies, such as flesh and bone, which could not be produced by the juxtaposition of these variously shaped elements, nor by their parts, the plane triangles.

The theory of the atomists was most suspect to Aristotle because they postulated the presence of a void in nature.[8] But beyond this, Aristotle argued, while the hypothesis of the constant movement of atoms might account for growth by apposition, it could not explain growing things having changed as a whole, either by admixture of something or by their own transformation. Further,

---

[6] Francis MacDonald Cornford, *Plato's Cosmology* (New York: Library of Liberal Arts, n.d.), p. 255.

[7] Richard McKeon, ed., *The Basic Works of Aristotle* (New York, 1941), "On the Heavens," Bk. III, ch. 8, 306b-307a.

[8] *Ibid.*, "On Generation and Corruption," Bk. I, ch. 8, 325a-325b; ch. 9, 326b-327a.

the atomist theory failed to account for certain types of alteration, for example, the transition of a body from the solid to the liquid state.

According to Aristotle, there must be an existing entity, a substratum, in which changes occur.[9] This was substance. Existing substance was composed necessarily of two constituents, matter and form, neither of which existed independently in the sublunar world. Pure matter was merely potentiality, the principle of individuality and the prerequisite of existence. Form contributed to substance the intrinsic qualities and external properties of individual bodies. Once matter existed in its lowest state, however, Aristotle asserted, it possessed inherent sources of change called "contraries" which were equated with the lowest types of form. The contraries were qualities, and they were four in number, "hot-dry," "hot-moist," "cold-dry," and "cold-moist." The simple bodies possessing each of the qualities were identified with the four elements of Empedocles, namely, fire air, earth, and water, respectively. Every compound natural body included some or all of these contraries in varying proportions. Aristotle attempted unsuccessfully to distinguish between what we would call at present chemical compounds and mixtures. He termed bodies "homoeomerous" if their division into ultimate minima—which was theoretically possible but practically impossible—showed no qualitative difference between the part and the whole.[10] Thus, among metallic substances, gold, silver, and tin were homoeomerous, as were flesh, bone, and fiber among animal and vegetable substances.

Aristotle did not develop a theory of minima in any detail. He made no reference to the size, shape, or properties of the ultimate particles. They merely embodied qualities or principles of change. He did, however, attempt to explain change in the chemical or physicochemical realm by reference to the presence or action of these elements within a body. Full maturity, that is, the realization of a thing's nature or purpose, or, in some instances, the realization of a latent form, was produced or was at least initiated by the thing's own natural heat operating on the passive characteristics within it.[11] Thus, those bodies that solidified as a result of heat,

---

[9] Ibid., "Physics," Bk. II, ch. 1, 192b-193b; "On Generation and Corruption," Bk. II, ch. 3-5, 330a-333a.

[10] A. D. P. Lee, trans., Aristotle Meteorologica (Cambridge, Mass., 1952), Bk. IV, ch. 10, 388a.

[11] Ibid., Bk. IV, ch. 2, 379b.

such as soda and salt, were composed of earth. Those that solidified as a result of cold, such as ice, snow, and hail, were composed of water. Also, there were some substances that solidified by the deprivation of both heat and moisture. These were composed of both water and earth prior to solidification, and mineral crystals were included in this category:

And some of these solids cannot be melted or softened like amber and some kinds of stone, for example stalactites in caves; for these too are formed in the same way, being solidified not by fire but because their heat is driven out by the cold and their moisture accompanies the heat when it retires.[12]

Further, there were some solids, for example iron, from which the moisture had not completely evaporated. These contained a preponderance of earth, but the presence of some water was indicated by the fact that they could be softened by heat. Here, of course, Aristotle documents the inability of his contemporaries to attain the fusion of some metals and minerals in their smelting processes, a condition which prevailed in Western civilization for a millenium and a half after the time of Aristotle.

In Aristotle's thought, then, there was a notion of a force that entered into the formation and cohesion of inorganic mineral matter. It was incorporated into the concept of active qualities operating on passive qualities and of the form of the body striving to attain its state of perfection. But the force was not dynamic; rather, it was developmental or organic in nature.

As S. Sambursky has shown, the Stoic concept of force was quite different.[13] In the Stoics' view, the pneuma, a mixture of fire and air, pervaded the cosmos, ruling and activating matter. It formed a mixture with the passive matter and prevented its disintegration. Each particular physical property of a substance—for example, the hardness of a stone, the transparency of a crystal, or the color of a gem—was thought to be the result of a definite mixture of the combined fire and air, and the totality of these properties or qualities in a body was called its *hexis*. The *hexis*, then, was the structure or the constitution of the inorganic substance. But these various pneumas which permeated the substance were conceived in dynamic terms; they were active and continuous through-

12 *Ibid.*, Bk. IV, ch. 10, 388*b*.
13 S. Sambursky, *Physics of the Stoics* (London, 1959), pp. 1-11, 21-23, 47-48, 86-87.

out the substance and quite possibly were thought of as wave motions in the continuous medium, simultaneously tensing the body and keeping it coherent. As against the qualitatively homogeneous atoms of Democritus and the homoeomerous particles of Aristotle, the Stoic conception of mixture also incorporated the belief that no two individual bodies could be the same; that is, they denied the principle of the identity of indiscernibles. Individuals differed in structure, no matter how alike they seemed in appearance and properties. Thus, the form of Aristotle was no longer the ultimate organizing principle of a body. It was replaced in the Stoic theory by a force continuum that united separated forms into passive matter and was the reason for the continued cohesion of the individual body.

These were the ideas forming the bases of theories concerning the nature of matter which arose in the seventeenth and eighteenth centuries. The hypotheses of the atomists, interwoven with those of Plato, to which the concept of particle attraction was added, became the molecular theory, while the Stoic emphasis upon the priority of force, together with Pythagorean notions, led in time to the postulation of the polar theory. And it was the Aristotelian theory, intermixed with animistic, mystical, astral, and alchemical ideas, which the adherents of both molecular and polar concepts successfully overthrew in the eighteenth century.

Alchemy had, as its theoretical basis, the teachings of Aristotle, that a mineral or a metal was a combination of matter and form, and that a mineral, by virtue of its own form or aided by nature, strove to perfect itself in the bowels of the earth. This concept was strengthened by the more animistic notions about minerals which were present in antiquity, that they were actually alive, procreated, and gave birth to their own kind. Theophrastus, Aristotle's successor as leader of the Peripatetics, mentioned—incredulously, it should be noted—that the greatest and most wonderful attribute of certain stones was their ability to give birth to young.[14] He did not mention the kinds of stones thus endowed, but Pliny the Elder (A.D. 23/24-79), the great encyclopedist of the early Roman Empire, contributed this information.[15] There were stones known as eaglestones, he reported, so called because they were found in the

---

[14] Earle R. Caley and John F. C. Richards, *Theophrastus on Stones* (Columbus, Ohio, 1956), pp. 46, 68-70.

[15] H. Rackham and W. H. S. Jones, trans., *Pliny: Natural History* (Cambridge, Mass.: Loeb Library, 1938), XXXVI, 149; XXXVII, 21, 32, 53, 100, 163.

nests of eagles who could not hatch out their young without them. Pliny listed four varieties and the localities where they could be found. They occurred in male and female pairs, and whereas the female carried inside it a pleasing white clay which was likely to crumble readily, the male stone was hard, colored like an oak gall or reddish in appearance, and contained a hard stone in its hollow center. Other stones called "gassinades," according to Pliny, betrayed the presence of another stone in their wombs, and the embryo took three months to develop. Hence, the belief in the power of minerals to generate was reinforced by this belief in "pregnant" stones.[16]

Alchemy, whose earliest adepts began writing recipes about 200 B.C. in Alexandria, also had a strong craft tradition among metalworkers and jewelers who worked with minerals and who were stimulated by the possibility of transforming the baser metals and semiprecious stones into the more expensive varieties. If one could separate the form from the prime matter and then recombine the matter with, for example, the form of gold, one would have duplicated the process that nature effected over a long period of time within the earth. Minerals, when converted into metals in the smelting process, released gases or "spirits," one of which could be recognized by its characteristic odor, that is, in modern terminology, sulfur in combination with oxygen. In this way mercury was recovered from its sulfide by roasting, and at about the beginning of the Christian era the Romans began to produce gold by treating its crushed ores with mercury, thereby forming a gold-mercury amalgam.[17] The amalgam was separated from the gangue by forcing the mixture through leather, and the gold was recovered by distilling off the mercury. In the minds of the alchemists then, pure mercury and sulphur—not the modern elements—became "spirits" or "principles," mercury the principle of metallicity and sulfur the principle of combustibility. The more sulfur a metal contained, the baser it was; the more mercury, the nobler it was. If the form could be stripped from the baser metals and the principle of metallicity added in the proper proportion, the difficult task of transmutation could be effected.

It is not the purpose here to dwell on the practices of alchemy

---

[16] For an excellent account of the various beliefs in the generation of stones, see Adams, *op. cit.*, pp. 77-136.

[17] Charles Singer *et al.*, ed., *A History of Technology* (5 vols.; Oxford, 1954-1958), II, 42.

nor on its mystical implications, but merely to emphasize that it was widely practiced during the early Christian era, that it flourished among Arabic physicians after the Islamic conquests, and that alchemical writings were translated and penetrated the Christian West in the later Middle Ages.[18] Astrological lore contributed to the belief that astral and planetary influences were at work in the concoction in the earth of the various minerals from which metals were extracted. The Neoplatonic doctrines that permeated Arabic thought strengthened the idea of panpsychism.[19] Hence this concept finds expression in western Europe from the thirteenth century on. The *De rerum principia* attributed to Duns Scotus (*ca.* 1270-1308) asserted that metals and stones have life.[20] In the writings of Paracelsus (1493-1541) there are clear statements of the belief in the vitality of stones and metals as well as of plants, animals, and men, although it is admitted that the organic structure of the particles of which these beings are composed differs.[21]

Despite a strong belief in panpsychism during the Renaissance, voices were raised against the idea that minerals possess life. Bernard Palissy (1510-1590), the confidant of Catherine de Medici, dismissed the notion that stones have vegetative souls and attributed their growth to what he termed congelative augmentation.[22] The most rational argument against the vegetative growth of minerals, however, came during the seventeenth century from Nicolaus Steno (1638-1686), one of the great natural philosophers of the period, who in his mature years abandoned science for a career in the Church. It is of significance to this question that Steno was not only a skilled anatomist but also a keen geological observer. As a medical student, he discovered the parotid duct, and during his work on the anatomy of the brain he found a pineal gland in animals like that of man, a result that contributed to the downfall of Descartes's unique theory of mind-body interaction. While occupying the position of court physician to Ferdinand II, Grand Duke of Tuscany, Steno studied the geology of the surrounding

---

[18] For a good survey of alchemy, see F. Sherwood Taylor, *The Alchemists* (New York, 1949).

[19] See, for example, the thought of Proclus in Thomas Whittaker, *The Neoplatonists* (Cambridge, 1928), p. 278.

[20] Thorndike, *op. cit.*, III, 7.

[21] Franz Hartmann, *Paracelsus* (London, 1896), pp. 99, 202.

[22] Anatole France, ed., *Les Œuvres de Bernard Palissy* (Paris, 1880), pp. 64, 68, 318-347.

countryside and formulated an essentially modern hypothesis of the deposition of geological strata and of the reasons for their changes. Steno's remarkable observations on the process of crystalline growth were contained in his short treatise, *De solido intra solidum naturaliter contento dissertationis prodromus* (Florence, 1669).[23] He explained that any animate or inanimate body grew because new particles secreted from an external fluid were added to the existing body. This accretion took place either immediately from an external fluid or through one or more mediating internal fluids. Drawing upon his anatomical knowledge, he explained in detail that plants and animals grow and are nourished by the process of intussusception, that is, by the agency of mediating internal fluids. Particles are added by one or more of these fluids, and they either become attached to an existing fiber or form a new fiber by deposition in an interstice of an existing fiber. In contrast, he asserted, the growth of mineral crystals takes place by the addition of particles directly from an external fluid. He admitted that he did not know why a first tiny seed crystal is produced, but he stated that after this occurrence growth takes place by the superimposition of particles on existing external faces. These are drawn to the crystal faces by the action of a subtle fluid within the existing crystal, whose action he likened to the direct action of a magnet on iron filings.

Steno's treatise was translated into English in 1671 by Henry Oldenburg, secretary of the Royal Society of London. The work undoubtedly exerted a powerful influence in discrediting the belief in the vegetative growth of minerals, as it was still being quoted as an authoritative source a century later.[24] But that the idea was tenacious is seen by its presence in eighteenth-century writings. Joseph Pitton de Tournefort (1656-1708), the most illustrious botanist in France at the turn of the eighteenth century, was an outspoken advocate of the theory that there were present in the earth mineral germs, endowed with an organizing principle, which drew in matter from the surroundings, ingested it, and then

---

[23] Modern editions are J. G. Winter, *The Prodromus of Nicolaus Steno's Dissertation Concerning a Solid Body Enclosed by the Process of Nature within a Solid* (New York, 1916), and Karl Mieleitner, trans., *Vorläufer einer Dissertation über feste Körper, die innerhalb anderer fester Körper von Natur aus eingeschlossen sind* (Leipzig, 1923).

[24] J. B. L. Romé de l'Isle, *Essai de cristallographie* (Paris, 1772), p. 168.

through veins deposited particles on outer surfaces, causing them to swell and show the phenomenon of growth.[25] In the *De la nature* (Amsterdam, 1761) of Jean Baptiste Robinet (1735-1820), which was thought at first to be the work of Diderot or Helvetius because of the relative obscurity and youth of its author, the central thesis was that nature is uniform in the treatment of all her productions, and consequently there must be a uniform generation of all beings in the animal, vegetable, and mineral kingdoms. Robinet asserted:

> It seems to me that one cannot give a more satisfactory answer other than to admit the existence of fossil germs, the development of which by an intussusception of matter yields minerals; new generations come into existence ceaselessly; stones engender stones, metals produce metals, as animals engender their kind, as plants produce plants, by seeds, germs, or eggs, because all these words are synonymous.[26]

Another partisan of this theory was Count Georges Leclerc Buffon (1707-1788). Although denying the idea of generation and growth by intussusception, Buffon, in his *Histoire naturelle des minéraux*, postulated the existence of an intermediate type of matter, partially inert and partially organic.[27] This matter was formed by the decay and putrefaction of animal and vegetable substances, but it possessed a kind of activity because it was composed of organic molecules. When this kind of matter, to which belonged calcinable substances and vegetable earths, combined with such inert matter as feldspar or basalt, with water serving as a vehicle, geometrically figured mineral crystals resulted. The dead organic matter conserved the active organic molecules which communicated to the passive matter traits of organization primarily characterized by a regularly shaped external form.

These theories, of course, stand in sharp contrast to the mechanistic thinking of the majority of eighteenth-century natural philosophers; yet they exemplify the persistence of ideas concerning nature. With the advent of the study of animal and vegetable physiology, they were seen to be erroneous, and in their grosser aspects they were forced outside the mainstream of scien-

---

[25] J. P. de Tournefort, "Description du labyrinthe de Candie, avec quelques observations sur l'accroissement et la génération des pierres," *ADS* (1702), pp. 217-234.

[26] J. B. Robinet, *De la nature* (Amsterdam, 1761), p. 290.

[27] Georges Leclerc, Comte de Buffon, *Histoire naturelle des minéraux* (5 vols.; Paris 1783-1788), I, 3-16.

tific thought.[28] Yet, panpsychism remained and, in fact, was encouraged by the German *Naturphilosophie* of the eighteenth and nineteenth centuries, which infiltrated thought both in England and the United States.

But the Aristotelian-alchemical legacy also pervaded the theoretical postulations of seventeenth- and eighteenth-century natural philosophers in more modest ways, and became modified as chemical knowledge increased. Particularly, there were present the idea that crystals were composed of concreted water and the belief that certain "principles" were operative which determined the characteristically geometric forms of crystals. Thus, in his *De Magnete*, William Gilbert (1546-1603) stated:

Lucid gems are made of water; just as Crystal which has been concreted from clear water, not always by a very great cold, as some used to judge, and by a very hard frost, but sometimes by a less severe one, the nature of the soil fashioning it, the humors or juices being shut up in definite cavities, in the way in which spars are produced in mines.[29]

We see that Gilbert believed the principal constituent of a rock crystal or other gems was water, but the crystallinity was not entirely attributed to cold; something in the soil aided its concretion. A half century later, Sir Thomas Browne (1605-1682), in his *Pseudodoxia Epidemica*, attempted to eradicate, along with other popular erroneous beliefs, this notion that rock crystal was permanently concreted water.[30] He argued that there were many authors and "exactest mineralogists" who denied it. Water ice was homogeneous, he pointed out, whereas he believed, mistakenly, that the substance of rock crystals contained mixed principles, mercury, sulfur, and salt—the elements of the alchemists—which were separable by the operations of chemistry. He referred to the differences in the specific weights of ice and rock crystal and called attention to the fact that crystals have electrical properties, that is, the power to attract straws when they are rubbed, whereas friction causes ice to melt. Turning to the shape of crystals, he

---

[28] Adams (*op. cit.*, p. 133) tells of a personal experience to demonstrate that the belief in the vegetative growth of stones is still present among the uneducated.

[29] William Gilbert, *On the Magnet* (New York, 1958), p. 51. First edition: *De Magnete* (London, 1600).

[30] Sir Thomas Browne, *Pseudodoxia Epidemica* (London, 1650), pp. 37-42. First edition: London, 1646.

stated that the majority were hexagonal, but that it was evident they did not receive their angularity by virtue of having been contained within an angular enclosure. By analogy, the stones sometimes found in the gallbladder of man had a triangular or pyramidal figure, although it was evident that the shape of that organ would not produce such a configuration. The shape of the crystal Browne attributed to a "seminal root," a formative geometric principle contained within it. He concluded that crystals were mineral bodies which had been liquid earthy matter. This fluid had hardened gradually owing to the coldness of the earth, and the final shape was determined by created and defined seeds within the fluid.

As chemical knowledge increased, there were attempts to link particular geometrizing principles to chemical properties. Francis Lana (1631-1687), for example, attributed the hexagonal figure of ice and snowflakes to short pointed nitrous particles which he believed entered into a type of combination with water.[31] Nehemiah Grew (1641-1712) held the same opinion.[32] He compared the figure of snowflakes with those of sal ammoniac, salt of urine, and feathers, wherein he saw a similarity. In an attempt to generalize the theory, he concluded that fowls having no organs for the evacuation of urine passed this material through the pores of the skin, where the urinous particles produced and nourished feathers. In a similar process, particles of falling rain met partly nitrous, but chiefly urinous, particles rising from the earth, and were fixed in their contact into the feathery form of snowflakes. Hence the organizing principle attached to these particular salts was featherlike in character.

More sophisticated was the late seventeenth-century view of chemists that the alchemical element, salt—a saline principle—was responsible for the solidity of substances, although many different types of salts had been distinguished. When crystals precipitated from a solution, the action was considered to have been due to the presence of salt, and the acid present in the solution was believed to cause the particular figure that resulted. Thus, Domenico Guglielmini (1655-1710) joined chemical properties to specific figures in his concept of the acid principle as the cause of a regular

---

[31] Francis Lana, "On the Formation of Crystals," PT, VII (1672), 4068.
[32] Nehemiah Grew, "On the Nature of Snow," PT, VIII (1673), 5193.

geometric crystalline form.[33] He believed that the first principles of common salt, the salt of vitriol, alum, and niter, were endowed with fixed and unalterable figures at creation. Primitive common salt was a cube, vitriol a rhombohedron, niter an hexagonal prism, and alum an octahedron. From these figures came those that they affected constantly in their crystallizations. When mixed with other kinds of matter, these primitive acid constituents combined to cause new figures which were modifications of the basic four.

Although a number of chemists supported Guglielmini's view, it was not without opposition. William Homberg (1652-1715) attempted to prove that an alkaline principle was responsible for the observed regular figures, pointing out that the same acid assumed different forms on crystallization depending upon the different alkalis or metals with which it was in solution.[34] The spirit of niter, which had dissolved salt of tartar, crystallized in long needles. If it had dissolved copper, the crystals were hexagonal, and if it had dissolved iron, irregular square crystals were produced. The crystalline figure of the same spirit of niter changed, then, according to the alkali with which it had crystallized. He observed that other acids behaved similarly, and he concluded that the figures belonged to the alkalis rather than to the acids.

Outside the laboratory, similar ideas prevailed in discussions concerning the forms of mineral crystals. Johann G. Wallerius (1709-1785) believed it was necessary to distinguish between the cause of crystallization and the cause of the figure of the crystals.[35] According to Wallerius, salt was unquestionably the cause of crystallization, but it appeared patent to him that mineral crystals had their origin from mixed earthy and metallic substances in which salt itself possessed no crystalline figure before it combined with something metallic. Thus he argued that the same metal dissolved in various acids took the same figure; for example, salts produced by the mixture of copper with nitric acid, sulfuric acid, or vinegar were always parallelepipedal crystals. Further, metals

---

[33] D. Guglielmini, *De salibus dissertatio epistolaris physico-medico-mechanica* (Venetiis, 1705). See also "Eloge de M. Guglielmini," *ADS* (1710), pp. 152-166.

[34] W. Homberg, "Essai de chimie," *ADS* (1702), pp. 44-45.

[35] Johann G. Wallerius, *Mineralogie* (Berlin, 1763), pp. 164-165. First edition: *Mineralogia eller mineral-Riket* (Stockholm, 1747). See also Elie Bertrand, *Dictionnaire universel des fossiles propres et des fossiles accidentels* (Avignon, 1763), p. 537.

themselves assumed certain definite figures; lead was almost always a cube, tin an elongated tetrahedron, and iron a rhombohedron or a cube. Hence, in Wallerius' opinion, the figure of a mineral crystal was due to its metallic ingredients.

But in some mineral substances no metal could be found. This was true of calcareous spars and diamonds, as William Borlase (1695-1762) pointed out. One therefore must have recourse to a saline principle, for " 'tis by the Force of Salts that liquid bodies are thrown into all the geometric Planes, Angles, and more compounded Shapes, the Variety of which is no less surprising than the Constancy and Uniformity of each particular species; the same salt shooting into the same figure."[36] The type of salt in the spars, Borlase believed, was most probably niter, which gave the ordinary hexagonal figure; when the spars formed other figures, it was because of the mixture of some other mineral or heterogeneous salt. The true diamond, Borlase thought, had little salt or water in its composition; it consisted almost entirely of "concreted lapideous juice," to which its property of extreme hardness could be attributed.

But the difficulties in maintaining the view that saline principles were responsible for all crystallizations and crystalline forms became more apparent as the methods of chemical analysis improved. John Hill (1716-1775) pointed out that some of the best minds were trapped by the illusions of the theory and that there was absolutely no proof of the existence of a saline principle.[37] The salts were bitter and soluble in water, whereas spars did not have any of these properties. Certain mineralogists, he asserted, had observed crystal forms of spars which were not paralleled in the figures of the ordinary known salts, and from these circumstances they had imagined that other salts must exist which were unknown to chemistry but which entered into such spars and might be recognized only because of this action. Such speculations, according to Hill, were useless; they created causes that were believed to be necessary for the explanation of observed effects. The Swedish chemist, Torbern Bergman (1735-1784), was in agreement

---

[36] William Borlase, "An Enquiry into the Original State and Properties of Spar and Sparry Productions, Particularly the Spars or Crystals Found in the Cornish Mines, Called the Cornish Diamonds," *PT*, XLVI (1749), 265.

[37] John Hill, "Spatogenèsie, ou traité de la nature et de la formation du spath," *JDP*, III (1774), 213.

with Hill in rejecting a saline principle.[38] As far as chemical analysis had been able to determine, he stated, certain gems and mineral crystals had nothing saline in them. Neither should the similarity of forms of various crystals be attributed to an acid principle; rather the external configuration of salts depended upon the joint combination of acid and alkali.

Involved in this conceptual difficulty was the inconceivability of the idea that some minerals might have been, prior to their crystallization, in a fluid condition at very high temperatures on the order of those attained by metals in a liquid state. In the seventeenth century, theories were presented by René Descartes (1595-1650), Athanasius Kircher (1602-1680), and others, similar to that popularized in the eighteenth century by Count Buffon, that the earth had at one time been a fiery mass that gradually cooled and eventually allowed habitation. These speculative hypotheses eventually were subsumed under the Vulcanist banner, which was opposed to the Neptunist theory.[39] The adherents of the latter postulated that the earth was covered originally by primeval waters carrying mineral constituents that gradually precipitated out in various stages as crystalline matter. The credibility of the idea that some minerals might have been fluid at high temperatures only was established when it was gradually realized that metals were crystalline substances and that the prerequisite of mineral crystallinity need not be an aqueous solution but also might be a state of high-temperature fusion.

As Cyril S. Smith has shown, it was René Antoine de Réaumur (1683-1757) who led the way in investigating the crystallinity of metals.[40] Réaumur, however, did not perceive the possibility of high-temperature fusion of minerals; rather he believed with Borlase and others that the formation of mineral crystals was effected by a crystalline essence or juice.[41] A contemporary of

[38] Torbern Bergman, *Physical and Chemical Essays*, trans. from Latin by Edmund Cullen (3 vols.; London, 1784), II, 24. First edition: *Opuscula Physica et Chemica* (Upsala, 1779).

[39] For an excellent account of this controversy, see C. C. Gillispie, *Genesis and Geology* (Cambridge, Mass., 1951).

[40] Cyril Stanley Smith, *A History of Metallography* (Chicago, 1960), pp. 102-111.

[41] R. A. de Réaumur, "De l'arrangement que prennent les parties des matières métalliques et minérales, lorsqu'après avoir été mises en fusion, elles viennent à se figer," *ADS* (1724), pp. 307-316; "Sur la nature et la formation

Réaumur, Johann F. Henkel (1670-1744), was, however, close to the idea when he stated: "But it cannot be conceived also, that metallic earths such as belong to the formation of ores, should be supposed to be dissolved by meer water, seeing we have neither example nor experiment to that purpose."[42] That metals had a crystalline structure was acknowledged from the 1720's onward; as we have seen, Wallerius referred to this fact. But this circumstance did not prove that minerals might once have been in a state of fusion. As J. B. L. Romé de l'Isle (1736-1790) expressed it in 1772,

To be sure, metallic substances that have been melted take certain determinate figures on cooling, but it is not necessary to interpose these deceptive appearances, nor to confound these rough cast figures with the true crystal forms which are produced by a slow operation of nature with the intermediation of water.[43]

Pierre C. Grignon (1723-1784) and Louis B. Guyton de Morveau (1737-1816) led the attack against this view. To Grignon, fire caused the fusion of metals, and slow cooling from the molten condition resulted in the production of crystals.[44] All types of crystallization, according to Guyton de Morveau, suppose a preceding solution.[45] Salts dissolve in water and crystallize because of the evaporation of the greater part of the fluid. In like manner, metals dissolve in fire, and they crystallize on cooling, that is, by the disappearance of the greater part of the solvent. Metals manifest equally with salts a regular arrangement of their composing molecules. Guyton de Morveau's examples were persuasive, but it was not until 1805, however, that the crystallization of minerals from a prior condition of high-temperature fusion was definitely established. In that year Sir James Hall reported to the Royal Society of Edinburgh the results of experiments with limestones which he had commenced as early as 1790. Upon being heated in a gastight gun barrel, powdered chalk was converted to a crystalline

---

des cailloux," *ADS* (1721), pp. 255-276; and "Sur la rondeur que semblent affecter certaines espèces de pierre et entre autres sur celle qu'affecte le caillou," *ADS* (1723), pp. 273-284.

[42] Johann F. Henkel, *Pyritologia, or a History of the Pyrites* (London, 1757), p. 254. First edition: *Pyritologia* (Leipzig, 1725).

[43] Romé de l'Isle, *op. cit.*, p. 322.

[44] P. C. Grignon, *Mémoires de physique: sur le fer* (Paris, 1775), pp. 476-481.

[45] L. B. Guyton de Morveau, *Elémens de chymie: théorique et pratique* (2 vols.; Dijon, 1778), I, 74.

mass of calcite.[46] His investigations, he stated, had been undertaken to confirm the Huttonian, or Vulcanist theory, of the formation of the earth's crust.

When the work of Antoine Laurent Lavoisier (1743-1794) and others in the closing decades of the eighteenth century put the science of chemistry on a firm basis by defining a chemical element in the modern sense, by isolating some of these elements, and by determining by quantitative methods the percentage of the elements in several substances, the Aristotelian-alchemical traditions with respect to crystals were given a death blow. The vague qualities, the four elements of the ancients, and the three elements of the alchemists were cast aside together with the saline and geometrizing principles. Two other theories relative to the structure of crystalline matter, however, had been emerging during the seventeenth and eighteenth centuries, the molecular, the most popular by far, and the polar. It is to the former that we shall first give our attention.

The molecular theory of crystalline matter was formulated by the merger of two ideas, first that crystals were formed by the juxtaposition of second-order particles, called corpuscles by some persons and molecules by others, which had definite geometric shapes that were not spherical, and second, that these molecules cohered owing to the action of an innate attractive force which they possessed. The first hypothesis was an outgrowth of those portions of the theories of the atomists and Plato which emphasized shape as one of the most important characteristics of the ultimate particles of matter, the atoms. But inasmuch as atoms were considered to be qualitatively homogeneous, it was difficult, except in a crude manner, to explain different physical and chemical properties by the characteristic of shape alone. Hence, some natural

---

[46] Sir James Hall, "Account of a Series of Experiments Shewing the Effects of Compression in Modifying the Action of Heat," *Philosophical Transactions of the Royal Society of Edinburgh*, VI (1812), 71-183. In an article entitled "On the Crystallizations Observed on Glass," *PT*, LXVI (1776), 539, James Keir suggested that "the great native crystals of *basaltes*, such as those which form the Giant's Causeway, or the pillars of Staffa, have been produced by the crystallization of a vitreous lava, rendered fluid by the fire of volcanos."

It should be noted that the use of the word "calcite" throughout this study is anachronistic, as it was first used by W. K. Haidinger in 1845. The English designated the substance "calcareous spar" and later "carbonate of lime." The Germans termed it "Kalkspath." The French called it "spath and later "chaux carbonatée."

philosophers, Descartes, for example, stressed particle size or differing degrees of particle motion, either innate or impressed, as
distinguishing features. Others emphasized that an infinite variety
of substances might result from different combinations of the
minima. Pierre Gassendi (1592-1655), one of the foremost adherents of atomism in seventeenth-century France, stated:

Hence, from atoms there are formed initially certain molecules, which
differ among themselves and which are the seeds of diverse substances,
and then each substance is constructed or composed from its own
seeds, so that nothing does or can exist from any other things.[47]

In another passage, Gassendi seized upon and extended the analogy
that Lucretius had made between atoms and the variety of substances that could be formed from them, on the one hand, and
letters as the elements of different words, phrases, and sentences,
on the other.[48]

It is Robert Boyle, however, who is generally recognized as
having been most influential in spreading the gospel of the corpuscular philosophy. Depending upon the ideas of Descartes and
Gassendi in many respects, Boyle believed there was one universal
extended matter, theoretically infinitely divisible, but in reality
divided into "prima naturalia," small particles having definite sizes
and shapes which had been set in motion at the creation of the
universe.[49] Aggregations of these particles were corpuscles or "primary clusters." In order to explain how these primary clusters
exhibited different physical and chemical properties, Boyle resorted to the idea that the shapes, order, and postures of the "prima
naturalia" might be modified to produce the characteristic "texture" of a substance:

. . . when many Corpuscles do so convene together as to compose any
distinct body, as a Stone or a Mettal, then from their other Accidents

---

[47] Petri Gassendi, "Syntagmatis philosophici," in *Opera Omnia* (Florence,
1722), I, 247. First edition: Lyon, 1658.

[48] A. G. van Melsen, *From Atomos to Atom* (New York, 1960), p. 92.

[49] Robert Boyle, *The Origine of Formes and Qualities* (2d ed.; Oxford,
1667). First edition: Oxford, 1666. This essay, which outlines Boyle's views
on matter, may be found in Thomas Birch, ed., *The Works of the Honourable Robert Boyle* (5 vols.; London, 1744), II, 451-487. For detailed accounts
of the development of the corpuscular philosophy, see Marie Boas, "The
Establishment of the Mechanical Philosophy," *Osiris*, X (1952), 412-541, and
Kurt Lasswitz, *Geschichte der Atomistik von Mittelalter bis Newton* (Hamburg, 1890).

(or Modes,) and from these last two mentioned [posture and order] there doth emerge a certain disposition or contrivance of parts in the whole, which we may call the Texture of it.[50]

According to Boyle, the primary clusters were not absolutely indivisible by nature, but they tended to remain so. They were the smallest units that could be detected by the analytical methods of chemistry. Boyle did not explain clearly how the cohesion of the diverse "prima naturalia" or the different corpuscles occurred. It appears that he tended toward a mechanical explication that was not so crude as that of Gassendi, who accepted Lucretius' notion of atoms with hooks and eyes. Boyle seemed to believe that some particles had pores that allowed the complete or partial entrance of other particles. This idea gained support from the experiments of one of Boyle's contemporaries, Nehemiah Grew.[51] The latter determined that water dissolved different weights of different salts and that the increase in the volume of water varied with the type of dissolved salt. From these phenomena, Grew concluded that the corpuscles of water were hard and inalterable in their figures, but that they did contain pores or vacuities into which the corpuscles of the dissolving salt might enter. But at the same time, the corpuscles of the various salts must have different figures, because, otherwise, the same number of porous particles of water would be able to imbibe as much of one salt as of another, and all salts would take up the same volume in solution. Perhaps this idea was similar to what Boyle had in mind; in any event, he did not emphasize any mechanical theory of cohesion.[52]

Agreeing with Nicolaus Steno, Boyle disliked and criticized the appeal to geometrizing or "architectonick" principles, as he termed them, to explain the regular angular forms of crystals.[53] He noted that some gems, probably referring to the garnet, contradicted this notion in that those that were dodecahedrons had rhombic surfaces rather than the pentagonal surfaces of the regular dodecahedrons of classical geometry. He was convinced that external, as well as

---

[50] Boyle, *op cit.*, p. 26.

[51] Nehemiah Grew, *The Anatomy of Plants* (London, 1682), pp. 297-300.

[52] See also John Freind, *Chymical Lectures* (London, 1712), p. 96, where calculations as to the relative sizes of the pore openings of gold and silver particles are made in an effort to explain why aqua regia will dissolve gold and not silver.

[53] Robert Boyle, *An Essay about the Origine and Virtues of Gems* (London, 1672), p. 55.

internal, conditions could influence the final crystal shape. As for external circumstances, he observed that the relative speed of evaporation had a marked effect. Boyle saw that "the Figures of these salts are not constantly in all respects the same, but may in diverse manners be somewhat varied, as they happen to be made to shoot more hastily or more leisurely, and as they shoot in a scanter or in a fuller proportion of Liquor."[54] Insofar as the internal conditions were concerned, he believed that the corpuscles merely chanced to adhere in one manner rather than another. The insensible corpuscles, he believed, all had "exquisite shapes, and [were] endowed with plain as well as smooth sides."[55] Boyle was uncertain whether or not the macroscopic crystal shape might give an indication of the shape of the composing corpuscles. He suggested that the corpuscles of niter might resemble the best-formed macroscopic crystals of that substance and that the characteristic taste might be due to the figures of the smaller fragments:

. . . if the corpuscles of nitre are prisms whose angles are too blunt or obtuse to make a vigorous impression on the tongue, yet if these prisms are broken by violent heat and made to grind on one another, they may have parts much smaller with sharp sides and angles, that being dissolved by the spittle, their smallness allows their entrance to the pores of the tongue and stabs.[56]

However, Boyle did not concentrate on this aspect; in another context he suggested that crystals might be formed of many thin layers or lamellae of the composing corpuscles. Perhaps he felt that it was useless speculation because the corpuscles were invisible. He did make use of the crystal form as a means of identifying particular salts; for example, he noted that the figure of alum was an octahedron, that of marine salt a cube, and that of saltpeter a prism.[57] Boyle was aware, though, that prudence should be observed in this identification. He pointed out that two salts whose shapes he knew could crystallize together into a mixed figure and thus confuse their species. He considered it a problem for geometers to explain how the mixed figure resulted from the simple forms of the individual salts.[58]

[54] Boyle, *Origine of Formes and Qualities*, p. 123.
[55] *Ibid.*, p. 126.
[56] Robert Boyle, "Mechanical Production of Tastes," in Birch, *op. cit.*, III, 587.
[57] Robert Boyle, "Determinate Nature of Effluviums," in *ibid.*, III, 329.
[58] Boyle, *Origine of Formes and Qualities*, p. 126.

With Gassendi and Boyle, Sir Isaac Newton (1642-1727) believed that there were ultimate minima that were small indivisible particles of several sizes and figures.[59] Their principal characteristics were solidity and hardness, which were the bases of the impenetrability of matter, and their innate power of mutual attraction and repulsion. Attraction caused the atoms to cohere into particles of the "first composition," and these in turn were attracted, though less strongly, to form particles of the "second composition." Newton concluded that the ultimate particles of matter were all transparent; the opacity of matter was due to internal reflections of light among the particles. He believed that sufficient evidence for the presence of interstices between the particles was presented by the change from opacity to transparency which took place, for example, when paper was soaked in oil, and the converse effect that occurred when the pores were removed, as when glass was powdered. He was not certain, however, whether these spaces were void or filled with a rare elastic fluid. The sizes of the component particles of a substance might be conjectured from their colors, because the transparent particles reflected light rays of one color according to their size and transmitted those of others. Color changes took place when liquids were mixed because the particles of one, in acting upon or uniting with those of the other, swelled or shrunk, thus changing the color of the rays that were reflected or transmitted. Newton believed that atoms might eventually be discovered with the aid of the microscope, but he believed that, because they were transparent, we would always be prevented from viewing their inner activity. In this manner, on the basis of optical theory, Newton postulated the sizes of corpuscles and some of their physical properties.

As with Boyle, however, we are left in doubt as to the shapes of the particles. Newton stated that they might have irregular forms, but it appears that he believed the particles of crystalline matter had regular geometric shapes. While remarking on the properties of crystals in general and the Iceland crystal, calcite, in particular, he said:

When any saline Liquor is evaporated to a Cuticle and let cool, the Salt concretes in regular Figures; which argues, that the Particles of the Salt before they concreted, floated in the Liquor at equal distances

---

[59] Sir Isaac Newton, *Opticks* (London, 1730; reprinted: New York, 1952), pp. 248-262.

in rank and file, and by consequence that they acted upon one another by some Power which at equal distances is equal, at unequal distances unequal. For by such a Power they will range themselves uniformly, and without it they will float irregularly, and come together as irregularly. And since the Particles of Island-Crystal act all the same way upon the Rays of Light for causing the unusual Refraction, may it not be supposed that in the Formation of this Crystal, the Particles not only ranged themselves in rank and file for concreting in regular Figures, but also by some kind of polar Virtue turned their homogeneal Sides the same way.[60]

When Newton treated the idea of particle attraction and repulsion, he transferred the concept of the attractive force of the planetary masses to the interaction of minute corpuscles, but for proof he depended almost entirely upon chemical phenomena.[61] The particles of acids, he thought, were much larger than those of water; consequently, they were endowed with a much greater attractive force and more activity. This superior force allowed acids to dissolve metals, excite heat, and destroy some particles by converting them to air (gas). Particles of salts, on the other hand, attracted those of water, and, in so doing, repulsed one another and diffused through the entire solution. In a sense, to Newton, attraction and repulsion were identical. He compared attraction passing into repulsion to numerical quantities passing from positive to negative. Repulsion could be considered a case of elective attraction; for example, when a particle was attracted more strongly by particle A than by particle B, it would repulse B in favor of A. He admitted that attraction seemed to be an occult property, but he felt that it was more reasonable to explain the cohesion of matter by a force than it was to resort to such crude notions as atoms with hooks or eyes, of relative rest between the particles, or by the pressure of an undetermined ether.[62] Newton's importance in the development of the science of crystallography, then, is due to his proposal of a reasonable explanation of cohesion of matter and to his belief that the corpuscles of crystals had regular geometric

---

[60] *Ibid.*, p. 388.

[61] The introduction to Volume II of John Harris, *Lexicon Technicum* (2 vols.; London, 1710) contains Newton's "De natura acidorum," in which these ideas are expressed. It is possible, however, that the sequence may be reversed, that is, that Newton arrived at the concept of gravitational attraction from his study of chemical affinity.

[62] Newton, *op. cit.*, pp. 388-389.

shapes. The hypothesis of attractive forces met with success when it was applied to explain the movement of the planets and the action of the tides. It gained in popularity, overcoming other theories of cohesion, and attained almost complete acceptance by the end of the eighteenth century.

Simultaneously, the belief grew that the tiny molecules of crystalline matter had regular geometric shapes and were juxtaposed to form the final crystal. Microscopical observations of crystalline matter were primarily responsible for the success of these ideas, but they also eliminated the necessity of assuming void intermediate spaces between the particles or interstices filled with an undetectable ether. It was for these reasons primarily that postulations that corpuscles were spherical, which were continually made in the seventeenth and eighteenth centuries, were rejected. But, in addition, the concept of spherical corpuscles presented the picture of matter of all chemical varieties, possessing diverse physical properties, being composed of qualitatively homogeneous spheres, an idea that ran counter to the emphasis that the corpuscular philosophy placed upon texture. Johann Kepler (1571-1630) attempted to explain the form of the snowflake by the packing of spherical particles in space.[63] Using this procedure, Kepler was able to demonstrate why cubic, tetragonal, hexahedral, and octahedral forms occurred, but because the snowflake exhibits hexagonal symmetry in a plane and because there are so many varieties of snowflakes, Kepler was unable to extend his construct to include them (fig. 4). In the end, Kepler resorted to the concept of a formative geometric principle at work in nature. Descartes, too, used the idea of spherical particles to account for the shape of snowflakes. He said:

And the small little clusters of ice ... are obliged to arrange themselves in such a way that each has six others surrounding it; one cannot conceive of any reason that would prevent their doing this, because all round and equal bodies that are moved in the same plane by the same kind of force naturally arrange themselves in this manner, as one can

[63] Johann Kepler, *Strena seu de nive sexangula* (Francofurti ad Moenum, 1611). See also R. Klug, *Des kaiserlichen Mathematikers Johannes Kepler Neujahrgeschenk oder über die Sechseckform des Schnees* (Linz, 1907), and a modern edition by H. Strunz and H. Borm (Regensburg, 1958), which I have not seen. For a penetrating discussion of Kepler's ideas see Cecil Schneer, "Kepler's New Year's Gift of a Snowflake," *Isis*, LI (1960), 539-544.

Fig. 4. Snowflakes. A few of the thousands photographed by W. A. Bentley of Jericho, Vermont. SOURCE: Snow Crystals, *by W. A. Bentley and W. J. Humphreys, p. 147. Copyright 1931 by McGraw-Hill Book Company. Used by permission of McGraw-Hill Book Company.*

see by an experiment, in throwing a row or two of completely round unstrung pearls confusedly on a plate, and shaking them, or only blowing against them slightly, so that they approach one another.[64]

In keeping with his mechanistic ideas, Descartes explained that the wind was the force that arranged the particles of the snowflake, and heat served to convert the equal round spheres of ice into miniature stars.

Robert Hooke (1635-1703) attempted to extend these ideas to include all crystalline substances by postulating that their particles were spherical, and by demonstrating that the regularity of crystal figures resulted from the configurations that contiguous spheres of equal diameters could assume. He stated:

... had I the time and opportunity, I could make probable, that all these regular Figures that are so conspicuously *various* and *curious*, and do so adorn and beautifie such multitudes of bodies, as I have above hinted, arise onely from three or four several positions or postures of *Globular* particles, and those the most plain, obvious, and necessary conjunctions of such figur'd particles that are possible, so that supposing such and such plain and obvious causes concurring the *coagulating particles* must necessarily compose a body of such a determinate regular *Figure*, and no other; and this with as much necessity and obviousness as a fluid body encompast with a *Heterogeneous* fluid must be protruded into a *Spherule* or *Globe*. And this I have ad *oculum* demonstrated with a company of bullets, and some few other very simple bodies; so that there was not any regular Figure, which I have hitherto met withall, of any of those bodies that I have above named, that I could not with the composition of bullets or globules, and one or two other bodies, imitate, even almost by shaking them together.

He continued:

And thus for instance may we find that the *Globular* bullets will of themselves, if put on an inclining plain so that they may run together, naturally run into a *triangular* order, composing all the variety of figures that can be imagin'd to be made out of *aequilateral triangles*; and such will you find, upon trial, all the surfaces of *Alum* to be compos'd of: For three bullets lying on a plain, as close to one another as they can compose an *aequilatero-triangular* form, as in the 7. *Scheme* [see fig. 5]. If a fourth be joyn'd to them on either side as closely as it can, they four compose the most regular Rhombus consisting of two *aequilateral triangles*, as B. If a fifth be joyn'd to them on either side in as close a

---

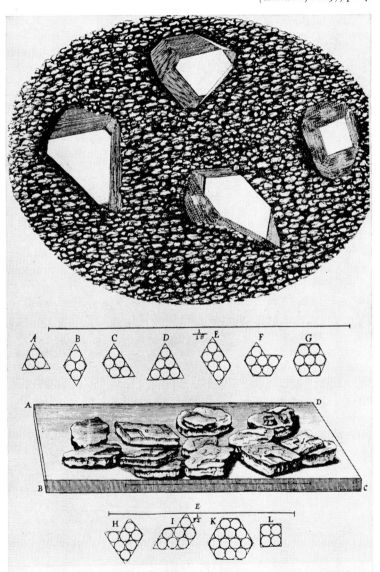

Fig. 5. Hooke's drawings of crystals observed under the microscope (above), and his postulated packing of spherical particles to account for the natural shapes observed in crystals of alum (below). SOURCE: *Robert Hooke,* Micrographia *(London, 1665), pl. 7.*

position as it can, which is the propriety of the *Texture*, it makes a *Trapezium*, or four-sided Figure, two of whose angles are 120. and two 60. degrees, as C. If a sixth be added, as before, either it makes an *aequilateral* triangle, as D, or a Rhomboeid, as E, or an *Hex-angular Figure*, as F, which is compos'd of two *primary Rhombes*. If a seventh be added, it makes either an *aequilatero-hexagonal* Figure, as G, or some kind of six-sided Figure, as H, or I. And though there be never so many placed together, they may be rang'd into some of these lately mentioned Figures, all the angles of which will be either 60. degrees, or 120. as the figure K, which is an *aequiangular-hexagonal* Figure is compounded of 12. *Globules*, or may be of 25, or 27, Or 36 or 42, etc. and by these kinds of texture or positions of globular bodies, may you find out all the variety of regular shapes, into which the smooth surfaces of *Alum* are formed, as upon examination any one may easily find; nor does it hold only in superficies, but in solidity also, for it's obvious that a fourth *Globule* laid upon the third in this texture, composes a regular *Tetrahedron*, which is a very usual Figure of the *Crystals* of *Alum*. And (to hasten) there is no one Figure into which *Alum* is observed to be crystallized, but may by this texture of *Globules* be imitated, and by no other.[65]

Hooke proceeded to explain that the figures of sea salt, sal-gem, vitriol, and saltpeter could be accounted for by his hypothesis, and he inferred that a subtle fluid probably filled the interstices between the spheres. Hooke's particles, then, were not qualitatively or quantitatively dissimilar spherical atoms grouped together in an orderly packing arrangement in space to constitute a crystalline configuration. They were chemically homogeneous spherically shaped molecules stacked together in different ways. He made no attempt to explain why similarly shaped molecules might have diverse chemical properties. His concept, then, was as useless as that of homogeneous atoms. Because of the growing proclivity to explain physical and chemical properties by reference to corpuscular texture, Hooke's hypothesis was outside the main current of scientific opinion.

Hooke reported that mica cleaved naturally into small rhombohedra, but he made no attempt to link this phenomenon to his spherical hypothesis.[66] Boyle, in thinking that crystals had a lamellar structure, attributed the directional character of crystal

---

[65] Robert Hooke, *Micrographia* (London, 1665), pp. 85-86. Also see the facsimile edition in R. T. Gunther, ed., *Early Science in Oxford* (14 vols.; Oxford, 1923-1945), XIII.

[66] Hooke, *op. cit.*, p. 48.

cleavage to a natural tendency for parting to occur parallel to these sedimentary layers.[67] In the seventeenth century, Christian Huygens (1629-1695) appears to have been alone in offering a different explanation.[68] In his study of the double refraction of light exhibited by the Iceland crystal, Huygens viewed crystal regularity as being the result of the ordering of similar invisible contiguous particles which were ellipsoids of rotation (fig. 6). His concept was, of course, quite similar to that of Hooke, and he may well have been led to it by Hooke's work. In his treatise, Huygens commented on Hooke's theory of light refraction which was included in the *Micrographia* along with Hooke's postulation of spherical particles. In Huygens' construct, each ellipsoidal particle

Fig. 6. Huygens' conception of the structure of the Iceland crystal. The small corpuscles composing it were conceived to be flattened spheroids such as would be produced by the rotation of the ellipse GH around its lesser diameter EF, the ratio of EF to GH being very nearly 1 to the square root of 8. The solid angle at point D would be equal to the obtuse angle of the crystal. Owing to the type of structure produced, Huygens explained that cleavage would take place readily on planes ADB, ADC, and BDC. Cleavage along planes marked GHKL, or ABC, although practicable, would be much more difficult. SOURCE: Christiaan Huygens, Treatise on Light, trans. from French by Silvanus P. Thompson (Chicago, 1912), pp. 96-97.

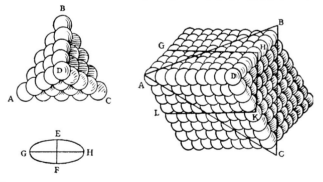

[67] Boyle, *Origine and Virtues of Gems*, pp. 22-23.

[68] Christiaan Huygens, *Treatise on Light*, trans. from French by Silvanus P. Thompson (Chicago, 1912), pp. 95-99. First edition: *Traité de la lumière* (Leyden, 1690).

of a cleavage layer was superimposed or touched on its flattened surface only one ellipsoid of an adjacent layer, being in point contact with two other ellipsoids of the layer. Contact on the flattened surface, Huygens supposed, was much more conducive to cohesion than was point contact of the ellipsoids. Cleavage along planes parallel to the rhombohedral surface of the crystal, then, required each tiny ellipsoid to separate from the flattened surface of only one adjacent ellipsoid of the contiguous layer. Other ways of division, he pointed out, would necessitate each ellipsoid detaching itself from the flattened surfaces of either three or four ellipsoids of the neighboring layer. At the present time, we understand that the planes of cleavage in a crystal are those across which the forces between the atoms are weakest. If a series of parallel atomic planes has weak binding forces between them, cleavage is likely to occur along these planes. The weakness may be due to the type of atomic bond or to a greater spacing of atoms or ions in the crystal at right angles to the plane of cleavage. By postulating a weaker bond between parallel cleavage planes, Huygens might appear to be a precursor of modern theory. To be sure, it is a much better representation than Boyle's sedimentary layers. But the deduction that Huygens made was a result of the construct itself. He denied the existence of attractive forces; he believed that cohesion was effected by the pressure of the ether. His model bears no similarity to the packing of calcium, carbon, and oxygen atoms in calcite, as we picture it today. We cannot, therefore, regard Huygens' hypothesis as an insight into modern theory. It was, of course, subject to the same criticism from the point of view of chemistry as was the hypothesis of Hooke. Huygens was fully aware of this fact when he stated:

I will not undertake to say anything touching the way in which so many corpuscles all equal and similar are generated, nor how they are set in such beautiful order; whether they are formed first and then assembled, or whether they arrange themselves thus in coming into being and as fast as they are produced. . . . To develop truths so recondite, there would be needed a knowledge of nature much greater than that which we now have.[69]

Hooke's microscopic observations of crystal forms were duplicated during the latter part of the seventeenth and throughout the

---

[69] *Ibid.*, p. 99.

eighteenth century. These investigations undoubtedly reinforced the idea that the molecules of crystalline matter were geometrically shaped particles, not spheres. A steady stream of crystalline substances passed under the microscope of Antony van Leeuwenhoek (1632-1723), and the *Philosophical Transactions* contain many reports of his findings, complete with sketches of the crystal shapes.[70] He described the figures of the particles of salt of wormwood, alum, saltpeter, salt of tartar, salt of English soda, salt of soda of Brittany, salt of alicant soda, sal ammoniac, and many others, all being geometrically regular in shape. In France, Antoine de Jussieu (1686-1758) was similarly occupied, not only confirming van Leeuwenhoek's findings, but also describing other substances. In England, Henry Baker (1698-1774) received the Copley medal in 1744 for his detailed microscopical descriptions and sketches of dozens of crystalline substances.[71] He wrote: ". . . we see every Species working on a different plan, producing Cubes, Rhombs, Pyramids, Pentagons, Hexagons, Octagons, or some other curious figures, peculiar to itself." The constancy of the figure of the crystals of the same substance proved, Baker asserted, that the component particles of the substance had some determinate and unalterable shape. In 1745, the memoir of Guillaume François Rouelle (1703-1770) describing the crystallization of common salt observed under the microscope also upheld this view.[72] When evaporation of the salt solution took place moderately slowly, Rouelle reported, one could instantly see, floating on the surface, tiny hollow pyramids whose points were truncated or square. But the small particles composing the pyramids had cubic forms themselves, and, as the evaporation of the solution continued and accretion of cubic particles occurred at the pyramidal surfaces, the pyramids gradually became cubes. If, however, the evaporation of the solution took place extremely slowly, crystals did not form at the surface but at the bottom of the receptacle and commenced as cubes from the start.

Thus the molecular theory of crystalline matter gradually gained

---

[70] See particularly Anthony van Leeuwenhoek, "Concerning the Various Figures of the Salts Contained in the Several Substances," *PT*, XV (1685), 1073; "Concerning the Figures of the Salts of Crystals," *PT*, XXIV (1705), 1906-1917.

[71] Henry Baker, *Employment for the Microscope* (London, 1764), pp. iv, 5-8. First edition: London, 1753.

[72] G. F. Rouelle, "Sur le sel marin," *ADS* (1745), pp. 57-79.

ascendancy, but it was not completely unopposed. Lancelot Law Whyte sees a sequence of six thinkers in the seventeenth and eighteenth centuries moving in parallel directions with respect to their theories of the constitution of matter: G. B. Vico (1668-1774), G. W. Leibniz (1646-1716), E. Swedenborg (1688-1772), R. J. Boscovich (1711-1787), John Michell (1724-1793), and Immanuel Kant (1724-1804).[73] Despite differences in the details of their theories, all rejected the concept of hard finite atoms extended in space and, instead, viewed matter as arising from the combination of centers of action or force which pervaded space. Whyte terms Boscovich's theory, *point atomism*, as contrasted to the *naïve atomism* of Robert Boyle and Sir Isaac Newton, expressions that correspond to my polar and molecular rubrics, respectively.[74] He views point atomism as originating in Pythagorean thought, according to which matter was postulated as developing from the movement of a point in space, first to form a line, then a plane, and ultimately a solid, with this idea being modified to include a force emanating from the points. Unquestionably, this Pythagorean notion is present, but, as S. Sambursky has argued, one should also take into account the Stoic influence on the development of seventeenth- and eighteenth-century theories of matter. Leibniz held the doctrine of a continuum and affirmed the principle that in nature there were not two like pieces of matter, both tenets of Stoic thought.[75] Further, there is an element of Neoplatonic thought in which the universe is seen as proceeding or developing from the infinite.

Upsala, at the time Emanuel Swedenborg studied there, was permeated by Cartesian natural philosophy.[76] Swedenborg early professed an anti-Cartesian bias, but despite this, unmistakably Cartesian elements, notably the vortex theory, remain in his works. That Swedenborg was familiar with at least portions of the Stoic

---

[73] Lancelot Law Whyte, ed., *Roger Joseph Boscovich* (London, 1961), pp. 118-119.

[74] *Ibid.*, pp. 106-107.

[75] C. I. Gerhardt, ed., *Die philosophischen Schriften von Gottfried Wilhelm Leibniz* (7 vols.; Berlin, 1885), VI, 608, 617. See Sambursky, *op. cit.*, p. 48.

[76] For this information about Swedenborg, see particularly Alfred Acton, ed., *The Letters and Memorials of Emanuel Swedenborg* (Bryn Athyn, Pa., 1948); C. O. Sigstedt, *The Swedenborg Epic* (London, 1952); and Signi Toksvig, *Emanuel Swedenborg* (New Haven, 1948).

theory of nature can be inferred from the fact that his doctoral dissertation concerned some works of the Roman Stoic, L. Annaeus Seneca (*ca.* 4 B.C. to A.D. 65). Further, there are distinct Stoic notions present in Swedenborg's *Principia*. For example, he likens nature to a spider's web:

If anything falls upon the threads or web, she, (the spider) lying in the centre as if in ambush, knows instantly where and in what part of the web it is. . . . Nature, then, is very much like this web. For it consists, as it were, of infinite radii proceeding from a centre, and of infinite circles or polygons, such that nothing can happen in one which is not instantly known at the centre, and thus spreads throughout much of the web. Thus through contiguity and connection does nature play her part.[77]

The same example of a spider's web, Sambursky shows, was used by the Stoics in antiquity to explain the principle of contiguous action "by the propagation of an impact through the medium of the taut threads of the cobweb."[78]

But Swedenborg's natural philosophy was eclectic. During his trip to England from 1709 to 1712, he read the works of Hooke, Boyle, and Newton, and he became acquainted with several of the English scientific luminaries. His theory of light was derived from Huygens'. He had studied the works of Leibniz and in 1714 visited Hanover hoping to meet this distinguished philosopher, only to find that Leibniz was in Vienna. Later, a correspondence developed between Swedenborg and Christian Woolf (1679-1754), whose philosophical work was a conscious attempt to systematize the thought of Leibniz. There seems to be no question that Swedenborg's ideas were influenced by those of Leibniz, whose philosophy of matter might be summed up in this excerpt from a letter to John Bernoulli: "And so it is possible, on the other hand—indeed, it is necessary—that there should be worlds not inferior to our own in beauty and variety, in the smallest bits of dust, in fact, in atoms."[79] Leibniz conceived that universal motion caused a progressive materialization of mathematical points, this process forming the ether

[77] James R. Rendell and I. Tansley, trans., *The Principia by Emanuel Swedenborg* (2 vols.; London, 1912), II, 541-542.

[78] Sambursky, *op. cit.*, p. 24.

[79] Leroy E. Loemker, trans. and ed., *Gottfried Wilhelm Leibniz: Philosophical Papers and Letters* (2 vols.; Chicago, 1956), II, 832. The letter to John Bernoulli is dated November 18, 1698.

and then spherical bodies that aggregated into the gross matter presented to our senses.[80] If the motions of the spherical particles were homogenous, the body was a solid; if they were hetero-geneous, it was fluid. If the heterogeneous motions of a fluid be-came unified and homogeneous, then the substance solidified. But there was little attempt at a rigorous explanation of the constitution of matter in Leibniz' natural philosophy; it was merely a minor appendage of his metaphysics.

In contrast, Swedenborg attempted to explain the formation of the entire universe, to account for all phenomena therein, to dem-onstrate how it worked mechanically, and to show it was subject to a universal mathematics.[81] He postulated a simple "first natural point" which proceeded out of the infinite and from which the finite universe was produced. This initial limit might be said to resemble a coin, one side of which looks toward the infinite and is therefore boundless, and the other side of which views the finite universe and defines it. Points of this type came into existence through the agency of God, and their only attribute was a *conatus* —a term borrowed from Leibniz—a striving toward spiral motion. Owing to the complex spiral motion of the point, there was pro-duced a surface, the "first finite," which was the smallest natural entity, and from which, in turn, everything in nature was gen-erated. Having an essentially spiral motion, the figure of the first finite had an equator and poles, and it could possess both directional and axillary motion. It was an "active" if it had both; mere axillary motion caused it to be classified as a "passive." From the motion of the first finites, second finites were formed which combined with the first to compose particles of the first universal element. In like manner, by the motion, compression, and expansion of the finites and elementary particles, a total of six types of finites, five elements, and some eleven kinds of particles were generated (fig. 7). These or their combinations allowed Swedenborg to explain the nature of the sun and stars, magnetic phenomena, interplanetary

---

[80] *Ibid.*, II, 837-879, particularly pp. 873-877 (correspondence with Burcher de Volder); II, 714-715 (Specimen Dynamicum). Also see J. A. Irving, "Leibniz' Theory of Matter," *Philosophy of Science*, III (1936), 208-214; E. C. Millington, "Theories of Cohesion in the Seventeenth Century" and "Studies in Capillarity and Cohesion in the Eighteenth Century," *Annals of Science*, V (1941-1947), 253-269, 352-369. For an eighteenth-century account, see M. M***, "Histoire des opinions philosophiques sur les principes et les élémens des corps," *JDP*, X (1777), 286.

[81] Rendell and Tansley, *op. cit.*, I, II, *passim*.

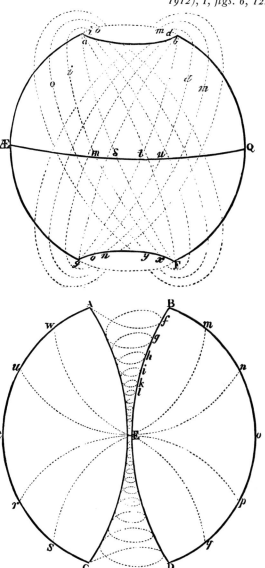

*Fig. 7. Swedenborg's illustration of an "active"
(above), showing the figure that its
movement describes and the circles by which
it forms its apparent surfaces. His
"elementary particle" (below) has polar cones.*
SOURCE: The Principia by Emanuel Swedenborg,
*trans. James R. Rendell and Isaiah Tansley
(2 vols.; London: The Swedenborg Society,
1912), I, figs. 6, 12.*

ether, air, and gross matter of an earthy nature. Concerning water, he said:

I would observe then that in every bubble of water there is contained all that had previously existed from the very first simple; every kind of finites, actives, and elementaries . . . so that in a small bubble the whole of our visible and invisible world is latent. We have thus the macrocosm in the microcosm; the world in a particle. . . . Nature is the same in the greatest things as she is in the least; in the whole as in the part.[82]

It would be tedious, even if possible, to detail Swedenborg's theory of matter, but it is necessary to emphasize that the poles of the various particles, particularly those composing the magnetic element, received a great deal of attention. The poles of the particles, are, in the main, pictured as vortices, created by the gyratory motion, into which can be drawn the tenuous matter within the boundaries of other proximate particles, the latter thereby losing their nature and changing into other types. Again, if at proper distances and aligned correctly, the suction action of the vortices of neighboring particles may cancel each other out, thus creating an equilibrium condition similar to Leibniz' concept of homogeneous particle motions resulting in a solid. Hence the polarities are viewed as being mechanical in nature, with the attempt made to explain cohesion mechanistically rather than by an occult innate force. The particles were of various sizes; they surrounded and interpenetrated one another. Since their surfaces were tenuous, arising only from the motion of points, they formed a continuum that could support vibratory or undulatory motion. Thus, too, according to Swedenborg, crystals were formed from spherical particles essentially, not those of Hooke or Huygens, to be sure, but they were packed together in the same orderly array.

Swedenborg may not have influenced Boscovich, but the latter's theory is of the same general nature.[83] The core of Boscovich's

[82] *Ibid.*, II, 266-267.

[83] There is, however, a considerable amount of literature on Swedenborg's possible influence upon Kant and Goethe. The difficulty appears to lie in separating out Swedenborg's mystical and scientific ideas. See Hans Schlieper, *Emanuel Swedenborgs System der Naturphilosophie* (Berlin, 1901); Jacques Roos, *Aspects littéraires du mysticisme philosophique et l'influence de Boehme et de Swedenborg au début du romantisme: William Blake, Novalis, Ballanche* (Strasbourg, 1951); Ernst Benz, "Swedenborg als geistiger Wegbahner der deutschen Idealismus und der deutschen Romantik," *Deutsche Vierteljahrschrift für Literaturwissenschaft und Geistesgeschichte*, XIX (1941), 1-32; and the biographies of Swedenborg listed above in note 76.

hypothesis is that the matter of the universe is composed of a finite number of nonextended points between which attractive and repulsive forces that are a function of distance only act in accordance with a simple law of force.[84] All the observed natural phenomena can be explained solely in terms of the distribution and motion of these points and the forces acting between them. Agreeing with Newton, Boscovich believed that the forces of attraction and repulsion were not different in kind; they differed only in direction in the same way that the successive subtraction of a quantity from a positive number results in a negative number. In order to illustrate the forces acting between two points, Boscovich constructed a graph, plotting the distance between the points along the abscissa, and the forces between them, as functions of distance, along the ordinate. Repulsive forces were manifested above the abscissa, and attractive forces beneath it. As the distance between the points approached zero, the curve of the law of forces approached the ordinate axis asymptotically on the repulsive side of the abscissa. Thus, two extremely proximate points were repelled from each other. At the other extreme, as the distance between two points approached infinity, the curve approached the abscissa as an asymptote on its attractive side. Thus a weak attractive force was present at great distances. Between these two extremes, however, the curve cut the abscissa frequently and was continuous, indicating the change from an attractive force to a repulsive force between two particles as a function of the distance between them.

Boscovich explained that the extension of matter arose from impenetrability inasmuch as several points could not occupy the same point in space at the same time owing to the forces of repulsion. But, with mutual forces interacting between all points in the universe, the state of any single point depended upon, to a certain extent, the state of all other points. Hence there was a physical continuum within which tenacious particles could be formed. These particles, not as immutable as the points, were termed particles of the first order, and, again reminiscent of Newton, particles of the second order could be constructed which were less tenacious than those of the first order. Cohesion of bodies was explained by the presence of the limit points on the curve of force, those points at which a passage from a repulsive force at a smaller distance to

---

[84] J. M. Child, trans., *R. J. Boscovich: A Theory of Natural Philosophy* (Chicago, 1922), pp. 83, 87, 89, 273-309. First edition: *Philosophiae naturalis theoria* (Venetia, 1755).

an attractive force at a greater distance occurred. Repulsion prevented the decrease, and attraction precluded the increase of the distance between the points, thus effecting a stable equilibrium.

Boscovich explained that there could be vast differences among the various groups of points which formed the particles of which bodies were composed. Two particles did not need to have equal masses, volumes, or densities, and there could be an infinite number of figures. The number and distribution of points forming the particles accounted for the great variety of substances and their properties. In some particles, an attractive force might prevail in the final configuration, in others, that of repulsion; still others might be neutral with respect to force. The difference between solids and fluids was that, although fluids did not lack mutual forces between the particles, these forces were not communicated to particles at a distance in any sensible degree, whereas there were lateral forces between the particles of solids. Moreover, bodies having parallelepipedal particles were harder than those formed of spherical particles because the surfaces fitted together more snugly.

Boscovich believed that his theory simply and satisfactorily explained the cause of crystallization and the regularity of the shapes of crystalline materials. He stated:

In particles such as are necessary for solidity, there is found quite easily the reason for a phenomenon pertaining to solid bodies, which is a source of the greatest wonder to physicists. That is, a disposition in certain special shapes which in salts especially seem to be quite constant; in ice, & the star-like flakes of snow more especially, they are wonderfully beautiful; & they observe certain definite laws, such as we also see, together with a constant shape of figure, in the simple constituents of crystals. But these are nowhere to be found so frequently as in the organic bodies of the vegetable and animal kingdoms. The reason for this comes out directly in this Theory of mine. For, if particles, at certain parts of their surfaces, attract other particles, & at other parts repel other particles, it can easily be understood why they should adhere to one another only in a certain manner of arrangement; that is to say, in such places only as there is attraction, & where there can be produced limit-points of sufficient strength; & thus, they can only group themselves together in figures of certain shapes. But since, in addition to this, the same particle, at the same part of its surface, with which it attracts one particle, will repel another particle situated differently with respect to it; whilst the mass of the great number of particles, set in motion at random, will slip by, those only will stay, which are attracted, and will adhere to those points in which, after approach, limit points of the greatest tenacity are produced. From this, the reason for secretion, nutrition, the growth of plants, & fixity of shape is perfectly evident.[85]

Here Boscovich expressed the idea that the symmetry evidenced in the organic realm of nature is attained by the identical process that is operative in the crystallization of inorganic matter.

It is not the purpose here to trace the flow of those ideas that were central to the polar concept of matter from one thinker to another or to follow their course through the eighteenth century. It is sufficient to say that the polar theory appeared as a principal tenet of German *Naturphilosophie* at the end of the eighteenth century, particularly in the thought of Friedrich Schelling (1775-1854).[86] Through Schelling it influenced the thinking of Christian Samuel Weiss (1780-1856), who applied it to the development of a new concept of crystal symmetry, one that set the science of crystallography on an entirely different course than that charted by René Just Haüy, an adherent of the molecular theory. These developments will be detailed in subsequent chapters. It is necessary first to look at another approach that was being followed in the effort to institute a science of crystals.

The twelfth edition of the *Systema naturae* (published in 1768) incorporated Carolus Linnaeus' (1707-1778) most mature thought concerning individuals of the mineral kingdom. There were, Linnaeus stated in the preface to this section, three methods one might use to investigate the stones, minerals and fossils found in the natural environment.[87] The first was the physical, which "descended through the obscure generation of minerals"; the second was the natural, which "considered the superficial and visible structures"; and the third was the chemical, which "ascended through destructive analysis." It was only wise and moderate, he remarked, to select the middle course, and, having thus provided a justification of his methodology, Linnaeus proceeded to make a systematic classification of various crystalline substances. We have reviewed the development of the physical and chemical theories of the nature of crystals. But the morphological approach was, in a sense, more typical of eighteenth-century science. It was a major influence in the development of the science of crystals, and it should therefore be considered in some detail.

---

[85] *Ibid.*, p. 311.
[86] Benz, *op. cit.*, p. 2, traces Swedenborg's thought to Schelling through Friedrich Christoph Oetinger, scholar and writer of Württemberg.
[87] Carolus Linnaeus, *Systema naturae* (Holmiae, 1768), III, 11.

CHAPTER III

The Classification
of Crystals

Modern philosophers of science, in the main, taking mathematical physics as the paradigm of satisfactory scientific explanation and prediction, regard classification, taxonomy, or natural history as merely descriptive, in essence a lower or meaner form of science, if such a method or approach to nature may be called scientific at all. For example, Edward H. Madden remarks about biological classification:

Linnaeus began the classification of the vast and ever increasing store of information on an effective basis according to genus and species; but this sort of activity, valuable and necessary as it is, did not make biology a science. . . . Biology, then, would be a science only if the biologist could explain why and how plants and animals come to have the characteristics they do, in fact, exhibit.[1]

Stephen Toulmin points out that "Natural historians . . . look for regularities of given forms, but physicists seek the form of given regularities." He continues:

In natural history, accordingly, the sheer accumulation of observations can have a value which in physics it never could have. This is one of the things which the sophisticated scientist holds against natural history; it is "mere bug-hunting"—a matter of collection rather than insight.[2]

---

[1] Edward H. Madden, *The Structure of Scientific Thought* (Boston, 1960), p. 113.
[2] Stephen Toulmin, *The Philosophy of Science* (New York, 1960), pp. 53-54.

That classification, however, is a necessary preliminary step to science, Ernest Nagel affirms:

The discovery and classification of kinds is an early but indispensable stage in the development of systematic knowledge; and all the sciences, including physics and chemistry, assume as well as continue to refine and modify distinctions with respect to kinds that have been initially recognized in common experience. Indeed, the development of comprehensive theoretical systems seems to be possible only after a preliminary classification of kinds has been achieved.[3]

Thus, it is suggested that mere classification, although a prerequisite for the development of science, is itself not science. At some point, perhaps when a sufficiently sophisticated system of classification has been erected by natural historians, true scientists can take over and transform—not just extend—the findings of their less gifted brethren by deriving rules or laws of nature which explain why the regularities occur as they do. Can such a generalization be supported? From a philosophical point of view, perhaps; historically, no. If one defines science as the explanation of observed regularities according to certain standards of accuracy and canons of logic, then one can, in fact, eliminate purely descriptive classifications as science in the defined sense; rather, they are systematic refinements of common-sense experience. But such a position discounts the thought, the insight, and the discrimination that are implicit in the development of any system of classification. It eliminates from serious consideration, as being in any sense scientific, the hypotheses that underlie the systematic tabulation of what has been observed. It neglects the correction of false starts, those systems based on hypotheses later found to be erroneous. It does not give sufficient recognition to the fact that a system of classification must necessarily be—in its essentials, at least—correctly structured in order for the explanatory theory or law of nature covering the particular regularity to be scientifically valid. Thus, from a historical point of view, one is assuming a very restrictive posture if one denies that classification is not only an essential but an integral part of the scientific enterprise. This fact is clearly apparent in the development of the science of crystallography.

Many difficulties beset those natural philosophers who attempted to construct systematic classifications of crystalline matter in the

---

[3] Ernest Nagel, *The Structure of Science* (New York, 1961), p. 31.

eighteenth century. Not only was there a lack of chemical knowledge in the modern sense, but also there was an absence of quantitative techniques in chemistry as well as in the measurement of such physical properties as hardness. These obstacles were reflected in a central problem concerning crystals. It was recognized that two or more kinds of matter which could be observationally differentiated by their chemical properties, by taste or activity with other reagents, by their physical properties, or by color or hardness, might crystallize with exactly similar external forms. But also it was noted that a substance that was by all known chemical and physical tests a distinct species or kind might still be found crystallized in two or more differing configurations. For example, calcareous spar, which in one locality might have a rhombohedral shape, in another would be found as an hexagonal prism, and, in still another, would have the shape of a scalenohedron. Given the similarity of form among diverse substances and the variation in shape in the same substance, was a classification system based on the crystal form feasible in the first place? How might the similarity, as well as the diversity, be explained in a reasonable manner? And if classification according to shape was not possible, what other characteristics of crystals might there be that could serve as the basis of a system.

Many natural philosophers of the first rank denied the possibility of crystal classification based on form. For example, Georg Ernst Stahl (1660-1734), the prominent German chemist, remarked early in the eighteenth century: "In general, with the exception of external identification, the figure of the salts does not appear to me to be of any utility; that is why I leave it to others to make discoveries of this sort."[4] And Count Buffon, writing in 1785, stated that "all the work of the crystallographers serves only to demonstrate that there is only variety everywhere where they suppose uniformity . . . that in nature there is nothing absolute, nothing perfectly regular."[5] Still, in the face of the seemingly paradoxical foundations and subject to the discouragement of their contemporaries, a number of scientists did make attempts to devise satisfactory systems of crystal classification. Beginnings had been made with this goal in mind in the course of the previous two centuries; we shall, however, consider here only the eighteenth-century work of Maurice

---

[4] Georg Ernst Stahl, *Traité des sels* (Paris, 1771), p. 284.
[5] Georges Leclerc, Comte de Buffon, *Histoire naturelle des minéraux* (5 vols.; Paris, 1783-1788), III, 433.

Cappeller (1685-1769), Linnaeus, A. G. Werner (1750-1817), and J. B. L. Romé de l'Isle. In addition, we shall trace the manner in which mathematics was introduced into the science of crystals.

All these systems of classification had as one of their principal roots the work of Nicolaus Steno, to which reference has been made in chapter ii in connection with the eradication of the concept of the vegetative growth of minerals. In his treatise, Steno also described the mechanism of crystal growth, using as an example, rock crystal or quartz, whose crystalline "habit"—the common and characteristic configuration in which a mineral crystallizes—is a hexagonal prism usually terminated by a combination of positive and negative rhombohedra, which are often so equally well developed that they simulate a hexagonal dipyramid.[6] Steno admitted that he did not know how or why the first tiny seed crystal is produced, but he stated that after this occurrence growth takes place by the superimposition of particles on the existing external faces of the crystal. The new matter of the crystal is not applied indifferently to all faces of the crystal but primarily to the surfaces of the pyramids, with the result that the size of the surfaces of the prism depends upon the sizes of the pyramidal faces. The prism faces, then, might be short or long, or sometimes absolutely lacking. Moreover, the particles were not always applied in the same quantity or at the same time to the faces of the pyramids. This circumstance might cause misalignment of the axes of the two opposed pyramids, inequality of the areas of the surfaces of the pyramids and prism, or absence of triangularity in the pyramidal surfaces. In some instances the new crystalline matter did not cover the entire surface to which it was being applied, so that the face became concave or convex, or protuberances like the steps of a staircase appeared on the crystal faces. Steno attributed this deposition process to the action of a subtle fluid emanating from the existing crystal, which he likened to the action of a magnet on iron filings.

Setting aside his appeal to a subtle fluid, this description by Steno of the mechanics of crystal growth is remarkable, particularly his assertion that growth is directional in character. He comprehended completely the fact that the eventual shape of the crystal depends entirely upon the growth rates in various directions. He would have understood the modern expression that a

<hr />

[6] J. G. Winter, *The Prodromus of Nicolaus Steno's Dissertation Concerning a Solid Body Enclosed by the Process of Nature within a Solid* (New York, 1916), pp. 237-245.

crystal is surrounded by those faces that are growing most slowly. He perceived that the rate of growth in any particular direction might vary with time. Scientists, at present, recognize that solution conditions, the degree of supersaturation at any face caused by density currents, or the presence of solution impurities which are selectively adsorbed at certain faces, modify the growth rates, although in some instances they cannot explain satisfactorily the latter factor. Steno's emphasis on the conditions of the solution in which the crystal grew was of importance in later thought, but more so was his remark that the modifications of the crystal form of hematite in the passage of time could be likened to a process in which the original form is truncated in certain ways at its solid angles and along its edges. We find this same idea in the work of Linnaeus and Werner, but it reaches its full extent with Romé de l'Isle. It became of prime importance in providing a rationale for crystal classification.

The Swiss physician, Maurice Cappeller, was familiar with Steno's work, as he was with the writings of some sixty other authors, who, until that time, had considered crystals from one point of view or another, and to whom he made reference. His classification of crystals appeared in a short treatise entitled *Prodromus crystallographiae, de crystallis improprie sic dictis commentarium* (Lucernae, 1723).[7] Although critical of Emanuel Swedenborg's work in some respects, Cappeller tended to agree with him that crystals were composed of tenuous particles, primarily spherical in shape, of water, ether, and salt. The particles of salt were constituted of a formless alkali and an acid which was primarily responsible for the crystal form. The acid of common salt yielded a cubic figure; vitriol gave a rhombic parallelepipedal form; niter formed triangular crystals; and alum crystallized as octahedrons. In crystallization, some of the water particles evaporated, allowing the union of the salt particles and causing groupings of these, the ether, and remaining water particles. Thus, the retained ether and water particles also contributed to the final shape of the crystal. But there were other contributing factors. The conditions under which crystallization occurred, as Steno had stressed, were also influential. Good crystal forms were produced in solutions lacking

---

[7] M. A. Cappeller, *Prodromus crystallographiae, de crystallis improprie sic dictis commentarium* (Lucernae, 1723). A modern edition, translated by Karl Mieleitner, was published in Munich in 1922.

mixed salt particles, wherein the liquid was undisturbed, and where crystallization proceeded very slowly. Crystals might also be formed by sublimation from mixed solutions at the surface of a liquid, as with ice, or by slow accretion, as with stalactites. Such solution conditions would not yield the more perfect forms obtained in the first instance.

Cappeller believed that nine types of forms would encompass all crystal shapes found in inorganic nature, for example, rounded, pyramidal, prismatic, polyhedral, and dendritic. Under each of the nine groups he listed all the stones, metals, earths, and salts that had been found in configurations resembling the class prototype. Forty drawings of crystals of a variety of substances which had been observed microscopically were appended to the treatise, and Cappeller emphasized that for proper classification it was necessary in some instances to employ the microscope. But, inasmuch as the manner of origin was conceded to influence the final shape, the same substance might be placed in several different classes and could theoretically appear in all.

Thus, Cappeller emphasized external form as a primary characteristic by which crystals could be classified. He felt that such a classification would be of practical use in medicine and pharmacy for the purposes of identification. But beyond this, he believed that it was an initial step in the stereometric investigation of the internal configurations of the particles composing crystals, which, together with the quantitative and qualitative analyses of their constituents, would bring to light the abstruse operations of nature. He said: "In a word, the complete form of the object may become recognized, which was said by Plato to be the object of science. We will understand the form, not in the abstract or metaphysically, but that which exists in determinate matter."[8] While admitting and even stressing that the crystals of a substance might show different configurations, Cappeller did not view this phenomenon as an insurmountable barrier to crystal classification. His work was received with great applause by his contemporaries, and, as a result of it, he was elected to membership in the Royal Society of London in 1726.

While in certain respects more precise and exhaustive than Cappeller, Linnaeus approached the problem of classification with

---

[8] *Ibid.*, p. 12.

less of a spirit of inquiry, demonstrating, instead, his adherence to a priori assumptions.[9] It was the similarity of the forms of different substances which impressed Linnaeus the most. He became convinced that he had found in the mineral kingdom the basis of a sexual system which had played so important a role in his botanical work. His primary classes were stones that he termed *steriles*, minerals that were *foecundi*, and fossils that were *ambigui*. The mineral class was divided into three orders, salts, sulfurs or inflammables, and metals. Salt, Linnaeus believed, or a group of saline principles, were responsible for all crystallizations. Niter, the aerial salt, "by obduction augments sand"; "muria," the marine salt, "by corrosion attracts clay"; "natrum," the animal salt, "by resudation coagulates calx"; and "alum," the vegetable salt, "by ramification cements soil." These salts, according to Linnaeus, were the fathers of stones, whereas clay, sand, soil, and calx were the mothers of stones. These primary saline principles were the sole causes of crystallization and had peculiar and determinate figures. To the genus niter he attributed the figure of a hexahedral prism; marine salt had a cubic figure; natrum had the shape of a tetragonal prism; and alum had an octahedral configuration. Varying numbers of mineral species were listed under these genera, each with a definite figure which could be obtained by truncating the edges or angles of the generic figure. For example, the amethyst and the diamond were species of the octahedral alum genus; the name given to the diamond was *alumen adamas*. In detailed descriptions, these subsidiary shapes were distinguished by the number of surfaces and edges they possessed and by the shapes of the surfaces. In these descriptions, Linnaeus explained how the figures resulted from the truncation of the angles and edges.

To a great extent, Linnaeus' work represents a transition between the old and the new. His ideas concerning the generation of minerals, though peculiar to himself, echoed the ancient beliefs of mineral procreation and vegetative growth without subscribing to them completely. He held fast to the conception of saline principles that had the capacity to organize substances geometrically, because they themselves were essentially geometric in nature. The form of the crystallized substance became the most important char-

---

[9] Carolus Linnaeus, *Systema naturae* (Holmiae, 1768), III, 3-7, 84-103, 213-216.

acteristic; chemical and physical properties were entirely sub-servient to it in his classification. But, at the same time, he made minute and accurate descriptions of the figures of many crystals and explained in what ways certain configurations were related to or differed from others. It was for this reason that both Romé de l'Isle and Abbé Haüy termed him the founder of the science of crystallography.[10] After Linnaeus' death, his mineralogical system was extended in Germany and in England.

With the spectacular extension of metalworking and metal-lurgical techniques in the rising industrialism of the eighteenth century, there was, of course, pressure on mineralogists to develop methods whereby minerals could be readily identified. It was owing primarily to this trend that Abraham Gottlob Werner, the renowned professor of mineralogy at Freiberg and the foremost protagonist of the Neptunist position, developed the science of oryctognosy, which was the determination of mineral species by painstaking examination and description of their external char-acteristics. Only one edition of his work, *Aeusserliche Kennzeichen der Fossilien* (Leipzig, 1774), was ever published, but it gained him widespread fame throughout Europe and America, and in his long tenure at Freiberg he developed the system in his miner-alogical lectures.[11] To a certain extent it is incorrect to speak of Werner's work as a mineralogical system; as Abbé Haüy put it, it was rather a system of mineralogical characteristics. Werner himself pointed out that the classification of minerals in a system and the identification of minerals from their external characteristics were two different things. He did not believe it was useful to apply rigidly the four graduated divisions of classes, orders, genera, and species to the inorganic realm, as Linnaeus had done.

Any mineral, Werner thought, had four essential characteristics: the external or sensible properties; the internal or chemical com-position; the physical, which included electrical and optical prop-erties; and the empirical, which described the place of its origin and its union with other minerals. While not rejecting the feasi-bility of mineral identification by one or more of the other char-acteristics, Werner believed the external characteristics were

---

[10] Haüy, *TDM2*, I, 9; J. B. L. Romé de l'Isle, *Essai de cristallographie* (Paris, 1772), p. 8.

[11] A. G. Werner, *On the External Characters of Minerals*, trans. Albert V. Carozzi (Urbana, Ill., 1962), p. xxiv.

"thoroughly complete, reliably discriminative, best known, easily recognized, [and] most convenient to determine."[12] The external characteristics were color, manner of cohesion, unctuosity or touch, coldness, weight, smell, and taste. The most important of these was the manner of cohesion, which incorporated external appearance, that is, the form, surface, and luster of the exterior; the internal appearance, or internal luster, type of fracture, and form of the fragments; and transparency, streak, hardness, solidity, and adhesion to the tongue. Thus, in Werner's system, the external form became only one of many different characteristics necessary to identify a mineral species. Werner explained that inasmuch as the external form was not precisely uniform, because it might be produced independently of the chemical composition owing to the effect of other circumstances, it could not be attributed the importance that it had been given in other systems of classification.

But Werner did teach that crystal forms might be differentiated according to whether they were primary forms or modifications of a primary form, and by size and manner of aggregation.

By the primary forms of crystals are understood their most simple forms, which consist of only one, or at most two, kinds of planes, namely lateral planes and terminal planes and which though they exist in crystals possessing several varieties of planes may be easily distinguished by conceiving those planes which lie nearest to the center of a crystal and which are generally the largest to be extended on all sides until they join.[13]

Werner believed at first that there were just six primary forms: (1) the icosahedron or polyhedron with twenty triangular equilateral faces; (2) the dodecahedron or the polyhedron with twelve pentagonal faces; (3) the prism, the number of whose sides was variable; (4) the pyramid, which might have differing numbers of faces and might also be double; (5) the table, which had two opposed faces larger than the others; and (6) the lens, which was terminated by two convex faces. To these he later added a seventh primary form, the cube.[14] Following Steno and Linnaeus, Werner emphasized that the

---

[12] *Ibid.*, p. 5.
[13] *Ibid.*, p. 55.
[14] R. J. Haüy, "Traité des caractères extérieurs des fossiles, traduit de l'allemand de M. A. G. Werner, par le tradacteur des Mémoires de chimie de Scheele, 1790," *ADC*, IX (1791), 174-192.

primary form might be modified by truncations of its edges and corners:

Galena has five kinds of crystals through which it passes from the perfect cube into the perfect double tetragonal pyramid. The perfect cube is its first crystal form; when its corners are slightly truncated, we obtain the second crystal form; when the truncation of the corners is carried farther until its sides are so large that they intersect each other, the third type of crystal form is obtained. . . . by truncating still farther we have the fifth and last crystal form of galena which is a perfect double tetragonal pyramid [octahedron].[15]

But, for Werner, these primary forms were in no way to be considered specific characteristics, that is, definite shapes in which certain minerals would crystallize, given optimum conditions of crystallization. They were merely simple idealized forms to which the configurations of crystals found in nature might be referred to as possible aids in identification. The perfect octahedron, he pointed out, might just as well be considered the primary crystal form of galena, and one might explain its transition into the perfect cube through the three crystal forms by reversing the method of truncation.

Thus, although crystal form played a role in Werner's system, it was far from being of outstanding importance. A certain range of variation was allowed in each of the various characteristics considered. If one mineral differed from another within the allowable range, it might be considered a subspecies of the other; if the range of variation was too large, the mineral was thought to be a different species. After the turn of the nineteenth century, when Abbé Haüy's system of mineralogical classification had gained prominence, Werner's work was attacked on the basis that the description of external characteristics could never raise mineralogy to the rank of a science, as Haüy was attempting to do, and that it was satisfactory only for the miner who was vitally interested in identification, or for the amateur.[16] But Werner had his champions also. Wernerian societies, designed to promote his ideas, were founded, the most prominent being in Edinburgh. It was admitted that the Wernerian system put emphasis upon utility, but it was

---

[15] Werner, *op. cit.*, p. 64.
[16] R. Chenevix, "Réflexions sur quelques méthodes minéralogiques," *ADC*, LXV (1808), 5-43, 113-160, 225-277.

argued that this should in no way detract from its scientific value.[17] The claim was made that his system was almost universally adopted and that, for this reason, Werner should be regarded as the founder of mineralogical science. For the science of crystallography, however, Werner's work was of importance because he restated the concept of primary forms, even though in his thought these were only ideals, and because he again drew attention to the fact that certain forms were related and could be derived from one another by the method of truncation.

It was Haüy's opinion that Werner, when he wrote his work in 1774, was not acquainted with Jean Baptiste Louis Romé de l'Isle's *Essai de Cristallographie*, published in Paris in 1772.[18] This is probable, but it is less probable that Werner's emphasis on truncation did not influence Romé de l'Isle in the writing of his second major work, *Cristallographie, ou description des formes propres à tous les corps du règne minéral*, a four-volume masterpiece which appeared in Paris in 1783. This work stands as a tremendous elaboration of the truncation concept, but with the very important difference that for Romé de l'Isle the primary forms were in no sense ideal or imaginary, but were real entities that existed in nature and to which the various mineral species conformed.

Romé de l'Isle appears as a somewhat tragic figure in the history of science. He was well educated and undoubtedly brilliant; he had a great interest in natural history, and his writings bear the mark of a keen mind and prodigious labor. They were widely acclaimed, and he was highly complimented by the great Linnaeus.[19] He was rewarded by membership in several foreign academies of science, including Stockholm, Mainz, and Berlin. But the major prize, the most distinguished mark of honor for any French scientist, membership in the Royal Academy of Sciences of Paris, eluded Romé de l'Isle. The expressly stated reason for this denial, that he was merely a pamphlet writer—a reference to the fact that he compiled several catalogs describing mineralogical collections that were to be auctioned—disguises the intrigue con-

---

[17] Thomas Thomson, "Some Observations in Answer to Mr. Chenevix's Attack upon Werner's Mineralogical Method," *AP*, I (1813), 241-258.

[18] Haüy, "Traité des caractères extérieurs des fossiles," p. 190.

[19] The letter from Linnaeus to Romé de l'Isle which characterizes the latter's *Essai de cristallographie* as the most distinguished mineralogical work of the century appears in J. B. L. Romé de l'Isle, *Cristallographie, ou description des formes propres à tous les corps du règne minéral* (4 vols.; Paris, 1783), I, xxi.

nected with the acquisition of a chair in the Academy.[20] It has been postulated that his spirited defense of the materialistic trend that was evident in France in his time caused his unpopularity.[21] Furthermore, his sharp attacks on the scientific theories of such academic luminaries as Buffon and Bailly could hardly have gained him influential friends.[22] In the end, he saw the prize go to Abbé Haüy, whose ideas and approach to crystallography he either would not or could not understand.

In prefacing his descriptive classification in his first treatise on crystallography, Romé de l'Isle stated that he was, to a great extent, following in the footsteps of Linnaeus. The figures of crystalline substances were found to be the same or almost the same in many diverse saline, mineral, and metallic substances; this similarity seemed to indicate a hidden affinity. He was an adherent of the molecular theory of crystallization, that crystals were formed by the juxtaposition of identical molecules of a constant and determinate figure, and he recognized that the degree of perfection of the resulting figure depended upon the purity and the amount of agitation of the solution from which the substance crystallized, as well as the length of time allowed in crystallization. Still he was inclined to attribute the figure of mineral crystals to a saline principle that entered into their composition, but he hesitated to embrace this theory completely. He agreed that pure metals were crystalline—though not so truly crystalline as minerals—and noted that no one had been able to extract a saline principle from them by chemical methods. He pointed out, however, that this failure did not prove that such a principle did not exist or might not be found in the future by improved analytical techniques. The most conclusive argument in favor of a saline principle in minerals, at least, was that stressed by Linnaeus: the striking similarity between such crystals as those of alum and diamonds. It was difficult for Romé de l'Isle to ignore this uniformity. He believed that the saline principle might be vitriolic acid, although he remarked that he did not wish to become involved in the controversy as to

---

[20] *Biographie universelle (Michaud)* (52 vols.; Paris, 1853-1866), XXXXVI, 406.

[21] This reason is given by Arthur Birembaut in Maurice Daumas, ed., *Histoire de la science* (Bruges, 1957), p. 1082.

[22] Romé de l'Isle ridiculed Buffon's theory of organic molecules in his *Cristallographie* (III, 573-575), and Buffon's and Bailly's theory concerning a central fire in the earth's interior in his short treatise *L'Action de feu central* (Paris, 1779).

whether all acids were derived from this primitive and universal substance.[23]

He defined a crystal as any body of the mineral kingdom which displayed a polyhedral and geometrical figure. He regarded the transparency or opacity of crystalline matter as an indifferent property. All crystals fell into one of four classes: (1) saline crystals, which were soluble in water; (2) stony or rocky crystals, which were relatively impervious to the action of fire; (3) sulfurous or arsenious crystals, which emitted disagreeable fumes when exposed to fire; and (4) metallic crystals, which became fluid by the action of fire.[24] These primary divisions, however simple or arbitrary they may seem, did have a logical basis conforming to the contemporary analytical methods of chemistry. The action of a substance when exposed to water or fire did serve as a rough indication of its chemical nature.

Romé de l'Isle's discussions of the nature of acid, basic, and neutral salts gives a good indication of the state of chemical knowledge in 1772. The problem, of course, involved the nature or the constituents of acids, and he explained acidity or alkalinity by reference to an elementary aqueous principle in the salt, which combined with some earthy element to produce either acidity, alkalinity, or an equilibrium between the two. Romé de l'Isle's primary purpose, however, was to list the forms of all salts produced by crystallization from mixtures of alkalies with each of the known acids. Each of the resulting salts was a separate species, and, from the varieties of shapes which each salt might take up on crystallization, he selected one he considered to be the most perfect figure. For example, he pointed out that the perfect form of alum was a regular octahedron whose faces were equilateral triangles. If truncations were made along the edges, variations from the octahedral shape would occur, and the resulting forms would correspond to all known configurations of alum crystals. In only a few instances did he note the values of the plane angles of the crystalline salts he listed, and in no instance did he give the value of any interfacial angle. He concentrated on description rather than measurement.[25]

As he had done with the saline crystals, Romé de l'Isle devoted space to explaining the theories of the origin and growth of stony crystals. He subdivided this class into spars, fusible spars, selenites,

---

[23] Romé de l'Isle, *Essai de cristallographie*, pp. 8-15, 34-45.
[24] *Ibid.*, pp. 6-7.
[25] *Ibid.*, pp. 59-61; Appendix, Pl. X.

gems, basalts, and zeolites, with mica and quartz as separate and individual species. Here a combination of chemical and physical properties served as the basis for the definition of each genus. For example, he characterized the fusible spars as differing from other stony crystals not only in their forms but also in such other properties as specific weight, hardness, phosphorescence, and fracture. The seven species of gems were diamond, ruby, sapphire, topaz, olivine, hyacinth, and emerald. Rarity, color, and brilliance were not, however, the only reasons for their inclusion in this category. All gems, he said, showed a marked difference in the disposition of their constituent particles. The gems were composed of small, hard platelets which were applied exactly on top of one another. This condition resulted in an intimate union of the particles and was responsible for the interior refraction of light and the brilliance of these stones. In contrast, micaceous crystals had a lamellar structure also, but their tissues were looser, coarser, and more friable.[26]

As had been the situation with salts, Romé d l'Isle described the perfect form of each of the forty-seven species of stony crystals which he considered, and listed the variations in figure to which each species was subject. The values of the plane angles of several species were noted and, in addition, the interfacial angles of rhombohedral selenite and of the Iceland crystal were given. It is doubtful, though, that at this time, 1772, Romé de l'Isle made his own angular measurements of these two minerals, because he referred to and used the exact angular values that Gabriel de la Hire (1677-1719) had reported in a memoir concerning these crystals, which he had written in 1710.[27] Also, the exact values of the angles seemed to be of little importance to Romé de l'Isle; for example, he reported the angles at the bases of the isosceles triangles of the hexagonal pyramids of quartz to be *about* 70° to 75°, and those of the summit angles to be *about* 30° to 40°.[28]

He admitted that sulfurous and arsenious crystals were akin to those of the metallic minerals, but he stated that they differed in form and specific weight, thereby justifying their separate status. This category comprised twelve species, the chief of which was

---

[26] *Ibid.*, pp. 148-195.
[27] G. de la Hire, "Observations sur une espèce de talc qu'on trouve communément proche de Paris au-dessus des bancs de pierre de plâtre," *ADS* (1710), pp. 341-352.
[28] Romé de l'Isle, *Essai de cristallographie*, pp. 115-176; Appendix, Pl. X.

pyrite. Mercury, antimony, zinc, bismuth, and cobalt were classified as demimetals. Of the true metals, tin, lead, iron, copper, and
silver ores were found in well-defined crystals, whereas gold and
platinum ores had no determinate polyhedral forms. He questioned
whether nickel, molybdenum, and manganese could be considered
metals, and he did not include them in his description of the crystals
of metallic minerals.

He added an appendix which included several plates showing
about a hundred sketches of crystal forms to which he referred in
order to clarify his textual descriptions. In addition, he included
a classification he called a crystallographic table, wherein 110 different forms were listed. The table described the figures of the
prism, if any, and the pyramids, if any, which composed each
crystal form, and gave the total number of its faces. It also
enumerated the names of salts, stones, and metallic minerals which
crystallized in a shape similar to the designated form, and, finally,
it correlated Romé de l'Isle's standard forms with those of Linnaeus. For example, Form No. 1 was composed of a hexagonal
prism with two hexagonal pyramids—one at each end of the prism—
which were shorter than the prism and had faces in the shape of
an isosceles triangle. This form, then, had a total of eighteen faces.
The salt that crystallized similar to this form was vitriolic tartar;
the stony crystals that took this figure were quartz, false hyacinth,
and amethyst; and the corresponding metallic mineral crystal was
a greenish lead mineral, possibly what we now call pyromorphite.
Finally, Form No. 1 in this classification was the same as that shown
in figure 1 of the twelfth edition of Linnaeus' *Systema naturae*.[29]

Romé de l'Isle's first effort toward the classification of crystals
did merit great praise. He described hundreds of various crystals,
and the practical benefits gained by the inclusion of the crystal
forms, as well as physical properties and other structural evidence,
as a means of identifying stones and minerals, were valuable. Thus,
he demonstrated how systematic description could be an important
addition to commonly used empirical tests. It is quite evident,
though, from his comments and from the structure of his crystallographic table that he believed the analogous crystal species of
salts, stones, and minerals depended upon a common principle, that
the same form indicated a basic similarity of some type in their
constitutions. Further, he was committed to the idea that the variations in form of each species did not result from the conditions of

---

[29] *Ibid.*, Appendix, Tableau cristallographique.

crystallization alone. Some other principle was responsible for these various modifications. He thought the concept of truncation, which could be applied abstractly in describing how one crystal form passed into another, was duplicated in nature by a principle that was somehow geometrical. But it is significant that, despite this conviction, he made no systematic measurements of any of the principal crystal angles, and it is apparent that he attached no particular importance to these values.

The statement of the crystallographic law, known as the law of the constancy of interfacial angles, is credited by some to Nicolaus Steno and by others to Romé de l'Isle. Because the internal structure of any crystalline substance is regularly repetitive and because the crystal faces bear a definite relation to that structure, it follows that the same faces have a definite angular relationship to each other. Steno noted in his treatise that in quartz, in the plane of the axis passing through the apexes of the opposed pyramids, the lengths of the faces of the pyramids and the prism might change without altering the angles created by their junction. Because of this observation, we must agree that Steno recognized the constancy of interfacial angles in quartz, but he did not generalize this statement to include other crystalline substances he studied. And he would have had to do this before his statement acquired any meaningful implication. The law was certainly recognized phenomenologically by a number of observers contemporary with Steno, but it seems to have made no particular impression, as is evidenced by its having been ignored in Romé de l'Isle's first work. In the first investigation of the double refraction of light in the Iceland crystal, published in 1669, the same year as Steno's treatise, Erasmus Bartholinus (1625-1698) noted the values of the plane and interfacial angles of the calcite rhombohedron.[30] But it is in Huygens' *Treatise on Light*, published in 1690, though communicated to the Academy of Sciences of Paris in 1678, that one clearly sees the law implicitly assumed but never explicitly stated. Huygens remarked that the same crystals were found in France, Corsica, and Iceland, thus indicating their complete identity. And he stated that he disagreed with Bartholinus slightly with respect to the values of the plane and interfacial angles, thus tacitly implying that the interfacial angles of the calcite rhombohedron are constant.

The most remarkable features of Huygens' treatise were his

---

[30] E. Bartholin, *Versuch mit dem doppeltbrechenden isländischen Kristall* (Leipzig, 1922). First edition: *Experimenta crystalli islandici . . .* (Hafniae, 1669).

concern with measurements, the accuracy of the values of the angles he noted, and the fact that his observations resulted in a clear conception of the axial symmetry of the crystal. He made exact measurements and calculations of the plane and interfacial angles, the angle of the deviation of the extraordinary refracted ray from the perpendicular, and the angles at which both faces and edges inclined to the axis of the crystal. He corrected Bartholinus' values of the angles of the rhombic surfaces from 101° and 79° to 101° 52′ and 78° 8′, respectively. He demonstrated that the most accurate method of measuring the interfacial angles was to drop perpendiculars from the blunt corner of the rhombohedron, which contains the three obtuse angles, to opposite edges; he reported these values to be 105° and 75°, a deviation of only 5′ from the modern values. We can grant that these and other data were gathered solely in order that Huygens might determine the value of the angle at which the faces inclined to the axis, which he calculated correctly to be 45° 20′, and which in turn was subservient to the determination of the ellipsoidal form of the surfaces of the light waves that he postulated. It was, however, the first accurate mathematical analysis of the form of any crystalline substance. Most important, though, was Huygens' recognition of the basic fact of crystal axial symmetry. He demonstrated this in making a pregnant suggestion, the importance of which was not recognized for well over a century.

Whence I concluded that one might form from this crystal solids similar to those which are its natural forms, which should produce, at all their surfaces, the same regular and irregular refractions as the natural surfaces, and which nevertheless would cleave in quite other ways, and not in directions parallel to any of their faces. . . . one would be able to fashion pyramids, having their base square, pentagonal, hexagonal, or with as many sides as one desired, all the surfaces of which should have the same refractions as the natural surfaces of the crystal, except the base, which will not refract the perpendicular ray. These surfaces will each make an angle of 45 degrees 20 minutes with the axis of the crystal, and the base will be a section perpendicular to the axis.[31]

In other words, the crystal axis was a directional property of primary importance; a ray of light impinging upon the crystal in the same direction as the crystal axis would not produce either

[31] Christiaan Huygens, *Treatise on Light*, trans. from French by Silvanus P. Thompson (Chicago, 1912), pp. 90-91.

ordinary or extraordinary refracted rays, whereas any one of a number of possible faces inclined to the axis at an angle of 45° 20′ should display the same optical properties as the natural faces of the calcite rhombohedron found in nature.

But just as his theory of identical ellipsoidal molecules composing the crystal fell on barren ground, so did Huygens' careful mathematical approach to the study of crystals. It appears most likely that in his *Opticks* Newton merely appropriated the angular values reported by Huygens and that he did not make independent measurements of these angles. Only a handful of authors measured the angles of crystals in the century separating Huygens and Romé de l'Isle, and these were confined to only a few substances. Why were such measurements not taken? Why was Huygens' example not followed, and why was the law of the constancy of interfacial angles not explicitly stated? It certainly cannot be attributed to the obscurity either of the author or of his work. I wish to suggest that it was owing to a conviction that if nature could be expressed in mathematical terms, then the mathematics must demonstrate a degree of simplicity. The lesson of Kepler's difficulty in the computation of the planetary orbits was overshadowed by the comparative simplicity of the laws he derived and of those of the Newtonian synthesis. Such findings as Boyle's law of the simple relation of the volume and pressure of air would serve to give additional support to this idea. What simple meaning, on the other hand, could be attached to such angular values as 101° 52′ or 45° 20′? It was not that Huygens' treatise was overlooked or ignored, but rather that the faith of scientists in mathematics as their most potent tool and as the language of scientific expression was not yet so firmly entrenched that they were prepared to accept the fact that the wording could be extremely complex.

In the early work of Romé de l'Isle, as has been noted, interfacial angles were observed on only two species, and their values were probably copied from the earlier work of de la Hire. Romé de l'Isle's recognition as the discoverer of the law of the constancy of interfacial angles resulted from his announcement in the preface of his 1783 edition that the respective inclination of the same faces was constant and invariable in each crystalline species.[32] This honor

---

32 *Cristallographie*, I, vi. Paul Groth, *Entwicklungsgeschichte der mineralogischen Wissenchaften* (Berlin, 1926), p. 7, credits Romé de l'Isle with the discovery, whereas Hélène Metzger, *La Genèse de la science des cristaux* (Paris, 1918), p. 67, states that Romé de l'Isle was the first to announce it.

has been misplaced or, at the very least, should be shared jointly between Romé de l'Isle and Arnould Carangeot (1742-1806), the inventor of the contact goniometer.[33] Carangeot was a student of Romé de l'Isle's at the time of his invention of this basic crystallographic measuring instrument. In the course of reporting his invention, Carangeot stated:

The author, a novice in crystallography, but very desirous of education in it by a profound study under the eyes and with the instruction of the creator of this science, was working at cleaving and modeling in clay life-size crystals which M. de l'Isle was to have done in terra-cotta to accompany the learned work that will appear shortly. Despairing after many tentative unfruitful attempts of not being able to render exactly a very bizarre form of rock crystal, he thought of cutting tentatively, out of cardboard, the angle that two of the faces formed with each other. When this angle had been cut, he was surprised to find the same angle on the two opposite faces, and so successively on the other faces of the same crystal. When the experiment was repeated on all rock crystals he had at hand, he recognized with satisfaction that the angles were constant, and produced, namely, 104° for the junction of the bases with each pyramid, and consequently 76° at their summit; 142° for the incidence of the faces of the pyramid on those of the prism; and 120° for each of the six angles of the prism, the form peculiar to that type of crystal. He hastened to announce the results of this experiment to M. de l'Isle. This savant, sensing its usefulness, had him repeat it, and recognized with the greatest pleasure that it took place constantly on the crystals of various mineral substances.[34]

Romé de l'Isle, then, did not recognize that the inclination of the faces was *exactly* the same in the same species until this fact was brought to his attention by Carangeot. It was necessary for Carangeot to repeat his work before Romé de l'Isle understood its implication. Carangeot, moreover, was not inclined to waive his claim to the discovery of the law; four years later he wrote: "As an example, I have given the rock crystal that occasioned my discovery of the property that is so interesting and peculiar to crystals, the constancy of the solid angles of those of the same species."[35] It follows logically that the law of the constancy of

---

Most modern textbooks and encyclopedias follow these authorities; see, for example, *Encyclopaedia Britannica* (1959), VI, 810.

[33] For a complete account of this discovery, see Daumas, *op. cit.*, pp. 1079-1082.

[34] A. Carangeot, "Goniomètre, ou mésure-angle," *JDP*, XXII (1783), 193.

[35] A. Carangeot, "Lettre de M. Carangeot à M. Kaestner," *JDP*, XXXI (1787), 204.

interfacial angles could not be declared absolutely until some measuring instrument, however primitive, was employed to assure that this phenomenon occurred in at least several species. Carangeot used the instrument and recognized the implications of his work. After his death it appears that only Haüy gave him credit for the joint discovery of the law.[36] It is noteworthy, too, that in stating the law Romé de l'Isle did not credit himself with its discovery, nor did he at any time deny, correct, or modify Carangeot's statements. Steel and copper goniometers accurate to within half a degree quickly appeared on the market, and in 1786 Carangeot reported an improvement of the instrument which allowed measurement of reentrant angles.[37]

The application of goniometry is one of the principal changes which is immediately apparent when the four-volume second edition of Romé de l'Isle's *Cristallographie* is compared with his earlier work. It is not only that the values of interfacial angles were given as a means of identifying individual species. Beyond this addition, geometry became the central focus of Romé de l'Isle's thought. He pointed out that Solomon appeared to have recognized a basic truth of creation when he wrote: "You ordered everything by measure and number and weight."[38] He quoted from Father Louis Bernard Castel (1688-1757) who in 1727 wrote that all aspects and parts of the universe were uniquely the object of mathematics. He agreed with P. C. Grignon that each body in the universe had a determinate and characteristic figure which, in conjunction with the chemical quality of the substance, determined its individuality. According to Grignon, matter could not exist without form; both came into existence at the same moment and together, simultaneously, endowed the body with individuality.[39] This seems to be the traditional Aristotelian definition, but to Grignon and Romé de l'Isle form was not that defined by Aristotle; instead, it was geometrical, capable of measurement, Platonic.

In attempting to delineate how the form characterized a substance, Romé de l'Isle revealed the upheavals that had taken place in chemical theory during the previous decade as well as the prob-

---

[36] Haüy, *TDM2*, I, 114.
[37] A. Carangeot, "Lettre à M. de Lametherie sur le goniomètre," *JDP*, XXIX (1786), 226.
[38] *Cristallographie*, I, 65. The quotation is from the apocryphal Wisdom of Solomon 11:21.
[39] *Ibid.*, I, 23n.

lems that still plagued chemists. He explained that the nature, number, and figure of the primitive elements were still unknown, but that fire and air appeared to be no longer indestructible. It was necessary, therefore, to deal with the secondary elements or the last results of chemical analysis, which were acid, phlogiston, an earthy principle, and an aqueous principle. It seemed justifiable to him to call phosphoric acid the universal acid from which all other acids were derived. Its universality, he thought, was proved by the effects of light and fire, by the decomposition and regeneration of air, and because it was common to both animals and vegetables, animal and vegetable acids being nothing more than phosphoric acid differently modified in these substances. If phosphoric acid were deprived of phlogiston, it became mephitic acid. The sole impediment to its claim of simplicity and universality was the difficulty in explaining how vitriolic acid was derived from it.[40] The contemporary confusion between the gaseous elements being discovered in increasing number at the time is quite evident in Romé de l'Isle's summary, and we can observe also the tenacity of the concept that there were just a few primitive principles to which all chemical phenomena might be referred.

Now, however, Romé de l'Isle proposed to demolish once and for all the idea that the figure of the molecules composing a body depended upon one or another of the chemical principles. The basic particles of a cube of common salt, he explained, were small cubes similar to the macroscopic crystal, but the primitive and constituent parts of this salt were acid and marine alkali, neither of which had a cubic form, although this form resulted from their combination with the intermediation of water. In like manner, the union of vitriolic acid and phlogiston yielded rhombic octahedrons of sulfur. Crystallization and the resulting crystal figure were intermingled inseparably, because a crystal was a polyhedral configuration resulting from the synthesis of the constituent principles of the body.[41]

Moreover, Romé de l'Isle no longer had recourse to a saline principle to account for crystallization. There were no substances in the mineral kingdom, he argued, neither salts, stones, minerals, nor metals, which did not submit to the law of crystallization. Crystallization was the necessary consequence of the great phenomenon

40 *Ibid.*, I, 8-13, 109-129.
41 *Ibid.*, I, 23-25.

of nature variously called impulsion, gravitation, weight, or attraction. He noted that some philosophers had postulated that all bodies in the universe, vegetable and animal as well as mineral, were in essence crystallizations, but he did not accept this hypothesis. He pointed out that an outstanding characteristic of the crystallizations of minerals is the production of bodies that are bounded by straight lines, whereas in plants and animals curved lines define the limits.[42]

The internal and hidden mechanism of crystallization, Romé de l'Isle said, is a mystery of nature, just as is the generation of animals and the vegetation of plants. When, however, the same chemical principles—acid, earth, the aqueous principle, or phlogiston—combined in the same proportions and under the same circumstances, crystalline bodies resulted which had the same form, density, and hardness.[43] Thus, form, density, and hardness delineated a crystal species, with form being the most important characteristic. They, in turn, depended upon the proportions of the constituent elements. Substances might crystallize perfectly, in which instance their shapes would be one or another of six primitive forms: regular tetrahedron, cube, regular octahedron, rhombohedron, rhombic octahedron, or dodecahedron with triangular faces.[44] Perfect crystallization resulted not only from ideal solution conditions, but also from a complete synthesis of the composing principles. A deviation from the former might produce a variation from the primitive form, but an incomplete synthesis caused a definite modification of the form, in essence, the production of a new species which was nevertheless related to the first. Romé de l'Isle stated: "The combination of two heterogeneous principles can be either perfect, as in all neutral salts, or imperfect with an excess of base or with an excess of acid, the primitive form being equally modified by these different degrees of saturation."[45] With an excess of base, the octahedral form of the crystals of alum, for example, was modified.

---

[42] Ibid., I, 18, 94. One outstanding proponent of the theory that crystallization was a universal phenomenon was Jean Claude de Lametherie (1743-1817). See his Principes de la philosophie naturelle (Paris, 1805), pp. 51, 138-161, 299. A short outline of the theory appears in J. C. de Lametherie, "Mémoire sur la cristallisation," JDP, XVII (1781), 251-265.

[43] Cristallographie, I, vii.

[44] Ibid., I, 74. For a further discussion of Romé de l'Isle's ideas on this subject, see R. Hooykas, "The Species Concept in Eighteenth-Century Mineralogy," Archives Internationales d'Histoire des Sciences, XXXI (1952), 45-55.

[45] Cristallographie, I, 73.

The substance remained almost the same, but the modified form, being a characteristic dependent upon the union of the constituent chemical principles, reflected a change in the proportions of these principles.

We can observe, by consideration of the six primitive forms, that there can be only one regular tetrahedron, cube, or regular octahedron, whereas countless representatives of the remaining forms may exist. Romé de l'Isle left unexplained the problem of why two otherwise entirely different substances might both have cubic, tetragonal, or octahedral forms. But he quickly seized upon the possible diversity in the other primitive forms as a means of distinguishing between two or more different substances. He explained that,

Although alum, sugar, and niter all have the rhombic octahedron as a primitive form, this octahedron has its faces differently inclined in the three salts. The octahedron of alum having the faces of its pyramids inclined at 55°, the obtuse angle formed by the intersection of the base of each pyramid is constantly 110°. This angle is 120° in the octahedron of niter, and 100° in the octahedron of sugar.[46]

Each crystalline substance, then, had a primitive form that was one of the designated six.

But how could this fact serve as the basis of a science of crystallography, and what would be the purpose and method of such a science? According to Romé de l'Isle, the regular and distinctive form of any substance whatsoever being known, and the values of its principal angles having been determined, it was easy to relate any known varieties to the primitive form by the employment of his theory and method of truncation. The science would consist in scrupulously describing all the primitive and accidental forms of crystalline matter and in specifying the more or less immediate relationships these forms have with one another. No reasonable science of mineralogy could exist without the science of crystallography, and the study of crystal forms, clarified by chemical analysis, should lead to real and satisfying discoveries in the field of mineralogy.[47]

This was the goal that Romé de l'Isle sought to attain. Meticulously, he demonstrated how secondary forms might be derived

<hr />

[46] *Ibid.*, I, 71.
[47] *Ibid.*, I, vi, 91.

from these primitive forms by truncating the latter either through the solid angles or the edges or both. The crystallographic tables which composed the fourth volume of his work included almost 450 variations of the six primitive forms. For example, there were sketches of 106 variations of the octahedron and 128 of the rhombohedron. Each form was carefully drawn, and the number of solid angles, edges, and faces of the form was noted. If a variation of one primitive form showed a close similarity to that of another, this fact was noted. The form of each salt, stone, and mineral described in the text could be found by reference to the tables. Thus, for example, thirty varieties of calcareous spar were listed. Romé de l'Isle admitted that the manner in which nature acted to simulate the process of truncation was an impenetrable mystery.[48]

Romé de l'Isle continued to use the primary divisions of salts, stones, pyrites, and metallic mineral crystals in his second edition, but now he remarked that this classification was arbitrary and superficial. He noted that some salts did not dissolve in water, whereas such stones as gypsum did. The forms were most important, but he did pay attention to the densities of substances and particularly to their hardnesses. Hardness, he believed, was the resistance of the composing molecules to mechanical division, and it seemed to be derived from the presence or relative proportions of the chemical principles that might inhere in the molecules. The presence of water, as in salts, would yield less hardness than that of stony crystals, in which acid acted with all of its energy at the time of formation. In metallic crystals, solidity and hardness appeared to proceed from a superabundance of phlogiston, for if this principle were removed by the action of fire, the metal became a powder. In passing, it should be remarked that this explanation bears mute witness to the magnitude of the task that Lavoisier, in the same decade, faced in his successful attempt to destroy the belief in phlogiston. Further, according to Romé de l'Isle, hardness generally depended not only upon the presence of these principles but also upon the different proportions of each which were present in a substance. Hardness, then, was far from being adventitious; it was as important as weight and form. Hardness, density, and form were the three specific characteristics of any substance, and he concluded that there did not exist in nature two substances intrinsi-

[48] *Ibid.*, I, xxvii.

cally different which had, simultaneously, the same crystalline form, density, and hardness.[49]

Romé de l'Isle believed he had arrived at a fundamental truth in his theory that there was a limited number of primitive crystalline forms in which variations might occur owing to solution conditions or which might undergo modification if the relative proportions of the composing chemical principles were altered. This was the hypothesis upon which he based his system of classification. It was founded not merely upon observation of superficial features of crystals; it was not just "bug-hunting"; instead, it was classification in accord with well-defined theoretical considerations. He even included in his crystallographic table some hypothetical forms that he was confident would be found represented in nature at some future time. In insisting upon the reality of just a few primitive forms, he introduced a concept that influenced the course of the science of crystallography for the next half century.

Nevertheless, though his contemporaries were impressed by the quantity of his observations and the accuracy of his descriptions of crystalline substances, they did raise objections to his method. First, there was the problem of the process of truncation. Although Romé de l'Isle emphasized that nature did not act in this way and that he was merely representing the effect of nature's processes by the theoretical truncation of the primitive forms so as to arrive at the secondary forms, critics argued that his ideas could not be sound because of this discrepancy. Some years later, one critic, in reviewing Romé de l'Isle's contributions in the light of Haüy's advances, stated:

Though often successful in explaining the origin of the most complex secondary forms by means of the imaginary truncations and bevellings of a simple solid, the immense industry and great sagacity of [Romé de l'Isle] were frequently baffled. . . . Though the hypothesis of *truncations* readily explains almost any appearance, accommodates itself with wonderful flexibility to difficulties, and introduces considerable facility into the expression of the most complicated crystalline forms, it is obviously inadmissible in any system that aims at an approximation to truth; because it involves an idea of diminution and subtraction, directly contrary to the most established principle of crystallization.[50]

---

[49] *Ibid.*, I, 15, 58-64; II, 364.
[50] Anonymous, "Review of M. Haüy's Traité de Minéralogie," *Edinburgh Review*, III (1803), 44, 46.

Further, it was clearly recognized that in the selection of primitive forms Romé de l'Isle had acted arbitrarily. One reviewer wrote:

De l'Isle, in declaring that the various forms observed in crystals of the same substance are only modifications of one constant primitive form, certainly announced a most important truth. It was a flash of genius; but in a philosophical enquiry, to prove it and not simply to say it, was the necessary step. . . . by a series of arbitrary truncations we may pass insensibly from any given form to any other. . . . As the combinations are infinite, a multitude of tables may be constructed. . . . the most simple of each table will be the primitive . . . nothing is proved.[51]

The criticism was just, and Werner recognized this fact in his consideration of truncated forms. Though Romé de l'Isle did not select the primitive forms capriciously, he justified their precedence by stating merely that they possessed a minimum number of faces. Beyond this, the frequency of occurrence undoubtedly entered into the selection of one particular form as primitive.

Finally, Romé de l'Isle gave little notice to the internal structure of a crystal. He could not overcome the conceptual difficulty that had plagued his predecessors in visualizing that the external shape was merely a manifestation of crystal growth; thus he could not arrive at the relatively sophisticated view that the external configuration still represented internal order. As a result, he attempted no explanation as to why the faces that resulted from the truncations had one inclination rather than another. He could not state with any certainty whether the hypothetical forms he included in his crystallographic tables could actually exist or whether they were necessarily excluded from crystallization. It was in just such questions that René Just Haüy was interested. Curiously enough, Haüy's work and goals were severely criticized by Romé de l'Isle in his *Cristallographie*. But before these rebukes can be analyzed, it is necessary to present an account of Haüy's approach.

---

[51] Abbé A. Q. Buée, "Outlines of the Mineralogical Systems of Romé de l'Isle and the Abbé Haüy: with Observations," *Nicholson's Journal*, IX (1804), 31-32.

CHAPTER IV

Haüy's Theory
of Crystal Structure

As noted in chapter ii, adherents of the molecular theory of crystalline matter in the latter half of the eighteenth century held that a crystal was composed of imperceptible second-order particles that were identically shaped and chemically homogeneous. Although some scientists still thought these molecules were joined by the exterior pressure of an ether, the theory that cohesion resulted from an attractive force residing in the particles was almost universally accepted by 1780. The molecular concept seemed vindicated by microscopic observations wherein the tiniest particles of crystalline substances were seen to have a variety of geometric shapes. From the second quarter of the eighteenth century onward, the hypothesis had been repeatedly advanced that the different forms might be accounted for by assuming that the tiny molecules were superimposed upon one another in a number of different ways. Although Romé de l'Isle accepted the molecular hypothesis and insisted that the variety of forms were related geometrically, his theory of truncation had not served to advance the juxtaposition concept. Establishment of a definite mathematical relationship between the variety of forms which a substance might assume upon crystallization was accomplished by Abbé René Just Haüy. At the basis of this achievement was a recognition of what was later called the crystallographic law of rational intercepts. Briefly, this law states that when referred to three intersecting axes all faces occurring on a crystal can be described by numerical indices which are integers, and that these integers are usually small numbers. The law was purely empirical, but the assumption of its universal applicability was fundamental in the mathematical approach to crystallography. Haüy's work in the development of the mathematical relationship between crystals of the same substance is thus of major importance in the progress of the science of crystals, and it is to the description of his work in this regard that the present chapter is devoted.

Haüy's successful synthesis of the prevalent ideas of crystal structure was brought to fruition primarily by his consideration of the cleavage phenomena of calcite. This interest, however, was in no way peculiar to Haüy. At the turn of the eighteenth century, Guglielmini had suggested that cleavage fragments represented the primitive polyhedra from which crystals were constructed. Years later, after some experimentation with the cleavage of calcite, Christian F. Westfeld (1746-1823) expressed the opinion in his *Mineralogische Abhandlungen* (Göttingen, 1767) that the crystals of calcite were formed of rhombohedral molecules and stated the problem with respect to its crystal varieties:

All crystals of spar are constituted of rhombohedral pieces, or rather, nature composed them actually in this manner; consequently, the principal cause of the formation is itself immaterial. It need only be asked why the rhombohedral pieces can assemble into crystals of another structure.[1]

An identical observation was made almost simultaneously by Johann G. Gahn (1745-1818), a student of Torbern Bergman.[2] The derivation of a definite form of calcite, the scalenohedron, from the cleavage rhombohedron was, however, made for the first time by Bergman himself, who published his findings in 1773.

According to Bergman, the calcite nucleus was a rhombohedron whose surfaces had angles of 101½° and 78½°. The various forms of calcite could be produced by superimposition of rhombic lamellae on each face of the rhombohedral nucleus. Gradually, as layers were built up, their sizes decreased according to some law, and various forms resulted depending upon the amount of decrease. Whether the actual decrease took place during the process of crys-

---

[1] See L. Sohncke, *Entwicklung einer Theorie der Kristallstruktur* (Leipzig, 1879), p. 8.
[2] Torbern Bergman, *Physical and Chemical Essays*, trans. from Latin by Edmund Cullen (3 vols.; London, 1784), II, 9n.

tallization, Bergman could not say, but the result, in any event, was a change in the appearance and number of the terminal faces of the crystal. With the help of careful sketches (fig. 8), Bergman

Fig. 8. Bergman's construction of a calcite prism (right) and the dodecahedron of garnet (left) from the juxtaposition of rhombic lamellae. SOURCE: Torbern Bergman, Physical and Chemical Essays, trans. from Latin by Edmund Cullen (3 vols.; London, 1784), II, pl. 1.

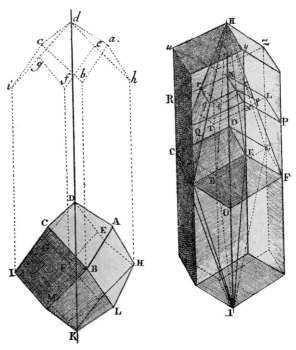

showed how that configuration of calcite, which consists of a hexagonal prism terminated at each end by rhombohedral faces, could be generated from a rhombohedral nucleus; he then successfully developed two scalenohedral forms by imagining that the superimposed layers continually decreased at their edges. Cleavage experiments on these varieties gave Bergman an indication that his hypothesis was correct. It a blow was aimed in a direction parallel to the assumed lamellar accumulations, cleavage readily took place.

Further, striations on the crystal faces gave evidence of the lamellar nature of the internal structure in the postulated directions. There was no question in Bergman's mind that his theory was capable of practical development: "If anyone imagines this doctrine to be purely geometrical and speculative, let him carefully examine the calcareous crystals, the loose texture of which, if cautiously and skillfully broken, will completely show the internal structure."[3] Bergman also considered garnet crystals and believed that the garnet nucleus was a rhombohedron also. He did not give the values of the facial angles of this rhombohedron, but one can infer, as Haüy did later, that Bergman erroneously believed they were the same as those of the calcite cleavage rhombohedron.[4] Nevertheless, he attempted to demonstrate that the usual dodecahedral configuration of garnet could be produced by successive additions of rhombic lamellae to this nucleus. In addition, he attempted to show how the peculiar cruciform figure of staurolite twins could be explained by imagining the addition of ever-decreasing layers to a nucleus consisting of a quadrangular prism terminated by quadrangular pyramids. Although he stated that he felt it unnecessary to multiply examples of this hypothesis, he confessed he was unable to derive by his method the hexagonal prism of calcite which had flat plane surfaces perpendicular to the axis of the prism at each end. He was at a loss to assign a reason for the disappearance of the apexes of such crystals. He believed that something unusual occurred because the truncated extremities were opaque, whereas the rest of the prism was transparent. He could only imagine that as more and more lamellae were deposited, they became progressively deficient around the axis, or that the rhombic lamellae were truncated across their faces in some manner, a supposition that violated the basic idea of growth by the deposition of rhombic lamellae.[5] Probably owing to the pressure of his exhaustive interest in chemistry and mineralogy, Bergman made no further attempts to derive crystal structure from cleavage fragments or to investigate the lawful relations between the various configurations of a substance. But, in concluding his analysis of crystal

---

[3] *Ibid.*, II, 9.
[4] R. J. Haüy, "Exposition de la théorie sur la structure des cristaux," *ADC*, XVII (1793), 243.
[5] Bergman, *op. cit.*, II, 10.

structure, Bergman made the statement that only more sagacious observers were required to bring the hypothesis to fruition.[6] He was correct; Haüy fulfilled his prediction within a decade.

René Just Haüy was born at Saint-Just-en-Chaussée (Oise) on February 28, 1743.[7] His father, a weaver, was apparently quite poor, and the boy received his early education at the nearby Premonstratensian abbey of Saint-Just. The family moved to Paris about 1750, and a few years later Haüy entered the University of Paris, receiving a scholarship at the College of Navarre. At that time the final year prior to the award of the Master of Arts degree included instruction in the sciences, in which Haüy was taught by Mathurin Brisson (1723-1806), a member of the Academy of Sciences and author of several works in natural history. Haüy was evidently an outstanding student because, after receiving his degree in 1761, he was appointed master of the third form at Navarre while he continued his theological studies. He rose gradually in the theological ranks, receiving his clerical tonsure in 1762, taking minor orders in 1765, becoming subdeacon in 1767 and deacon in 1769, and being ordained a priest in 1770. At this time he became master of the second form at the College of Cardinal le Moine.

During this period he developed an interest in botany, which continued throughout his life, although it remained on an amateur level. A letter to Pierre-Joseph Macquer (1718-1784), then professor of chemistry at the Jardin du Roi, testifies also to Haüy's interest in experimental chemistry. The turning point in Haüy's career, however, was his attendance at a series of lectures in mineralogy at the Jardin du Roi given by Louis J. M. Daubenton (1716?-1800). Haüy, fascinated by the subject, started his own collection of minerals and spent his spare time visiting and studying various mineralogical cabinets in Paris.

    ⁶ *Ibid.*, II, 23.

    ⁷ The biographical information on Haüy is derived from the following sources: *Biographie universelle* (*Michaud*) (52 vols.; Paris, 1853-1866), XVIII, 574-582; G. Cuvier, *Recueil des éloges historiques* (Paris, 1861), II, 255-284; Joseph Bertrand, *L'Académie des sciences et les académiciens de 1666 à 1793* (Paris, 1869), pp. 397-399; George F. Kunz, "The Life and Work of Haüy," *American Mineralogist*, III (1918), 61-89; Herbert P. Whitlock, "René Just Haüy and His Influence," *ibid.*, pp. 92-98; Alfred Lacroix, "La vie et l'œuvre de l'abbé René Just Haüy," *Bulletin de la Société Française de Minéralogie*, LXVII (1944), 15-94; and Maurice Daumas, ed., *Histoire de la science* (Bruges, 1957), pp. 1083-1088.

Haüy's biographers do not report the dates of the period when his interest in mineralogy blossomed. Almost all, however, from Georges Cuvier, who delivered Haüy's eulogy onward to the mid-twentieth century, repeat the details of an event in Haüy's life which parallels the story of the fall of the apple that supposedly gave Newton his insight into the theory of gravitational attraction.[8] While examining one of these mineralogical collections, Haüy is reported to have dropped and broken a group of calcite that had crystallized in the shape of hexagonal prisms. As he stooped to pick up the debris, he noticed that the fragments were rhombohedra, corresponding in every detail to the shape of the Iceland crystal. Immediately the idea came to him that such rhombohedra must be the nuclei of all calcite crystals. In a state of excitement, he returned home and broke two additional crystals of calcite from his own collection in order to validate his hypothesis. Both pieces, one crystallized as a hexagonal pyramid and the other as an acute rhombohedron, yielded the rhombohedral fragments he had observed previously. At this juncture, as Cuvier reported it, the abbé, without hesitation, cried out: "Tout est trouvé!" The event may certainly have happened, but, as has been noted, the idea that the cleavage nucleus of calcite was a rhombohedron had been stressed repeatedly since 1767, at least ten years before the occurrence.

Bergman's treatise appeared in 1773 in Latin, a language with which Haüy was quite familiar, and his collected works commenced publication in 1779.[9] In his later years, Haüy stated merely that he had arrived at his theory by the study of a fractured calcite crystal given to him by M. de France du Croisset at whose residence Cuvier placed the accident. Haüy wrote on several occasions that he was completely unacquainted with Bergman's successful derivation of the calcite scalenohedron from the cleavage rhombohedron until after he had informed the Academy of Sciences of the substance of his first two memoirs in 1781.[10] If so, it is a strange coincidence that Haüy's first memoir was concerned with the structure of garnet crystals while the second treated the crystal forms of calcite which had interested Bergman.

---

[8] All the biographers listed in note 7, except Arthur Birembaut in Daumas, *op. cit.*, repeat the story. It has thus been picked up by the encyclopedias, e.g., *Encyclopaedia Britannica* (1959), XI, 259.
[9] Bergman's treatise, "Variae crystallorum formae a spata ortae," was published in *Nov. Acta Reg. Soc. Sci. Upsal.*, I (1773).
[10] Haüy, *TDM2*, I, 15n; see also Haüy, *TDC*, I, 32.

Recently, R. Hooykas has concluded that Haüy was entirely in Bergman's debt and that the supposed breakage of calcite never occurred.[11] In his analysis Hooykas notes that in his first two memoirs Haüy remained on the same qualitative level as Bergman. In these Haüy explained the development of a crystal variety from the postulated nucleus by the superimposition of successively decreasing two-dimensional lamellae. Such a theory would, of course, prevent him from applying the laws of decrement by polyhedral molecules, laws that he subsequently postulated. There is no mention of laws of decrement in his 1781 memoirs. It is reasonable to infer, then, that Haüy was influenced in his initial work by Bergman, but thereafter he arrived at the idea that the lamellae were composed of cleavage rhombohedra and developed the laws of decrement between 1781 and 1783. It does not detract from Haüy's stature as the most influential pioneer of the science of crystallography to question the assertion that he was completely independent of the contemporary ideas of crystal structure. He accomplished what Bergman failed to do: he derived laws by reference to which the variety of the crystal forms of a substance might be explained.

With the encouragement of Daubenton, Haüy presented his first two memoirs to the Academy of Sciences on January 10 and December 22, 1781. This work and Haüy's continued diligent study of other crystalline substances impressed the members to such an extent that he was named an associate in the botanical class on February 8, 1783. Haüy's initial treatise, *Essai d'une théorie sur la structure des cristaux*, was published in 1784, and it is in this work that the laws of decrement were first successfully demonstrated. In this year also Haüy became emeritus professor so that he could devote all his time to the extension of his theory. From time to time he gave informal lectures to academicians and students who wished to understand the details of his work. The great naturalist, Etienne Geoffroy Saint-Hilaire (1772-1844), then a student at the College of Cardinal le Moine, recorded the attendance and demeanor at one lecture, in March, 1792, of Joseph Louis Lagrange (1736-1813), Pierre Simon Laplace (1749-1827), Claude

Louis Berthollet (1748-1822), Antoine François Fourcroy (1755-1809), Lavoisier, and Guyton de Morveau, a roll of the most distinguished contemporary French scientists.

From 1784 until his death in 1822, Haüy published more than one hundred memoirs in which he generalized his theory and analyzed the structure of scores of crystalline substances. He also investigated in detail the phenomena of pyroelectricity and double refraction in crystals. His books included a work on electricity and magnetism in 1787, an abridged edition of his theory of crystals in 1793, a four-volume work on mineralogy in 1801, a two-volume treatise on physics in 1803, and a mineralogical classification in 1809 which compared the results of chemical and crystallographic analysis. Two later editions of his physics work appeared, and, in the year of his death, he published a two-volume second edition of his treatise of crystallography and a four-volume second edition of his mineralogical work. His style was clear and concise, and he relied on copious illustrative sketches to ensure the comprehension of his presentation.

During the early years of the revolution, Haüy worked on one of the committees of the Academy of Sciences which was seeking to establish bases for the metric system. He refused, however, to take the oath required of all clerics by the Civil Constitution of the Clergy, and, after the overthrow of the monarchy on August 10, 1792, Haüy was among those members of the clergy arrested as suspect counterrevolutionaries. Through the efforts of Geoffroy Saint-Hilaire, he was released after a few days' confinement. At this time he accepted the civil certificate Geoffroy Saint-Hilaire procured for him and thus escaped the September massacres of the clergy which occurred at the Saint-Firmin seminary.

After the dissolution of the Academy of Sciences in August, 1793, Haüy became secretary of the temporary commission on weights and measures. In the following year he became associated with the Agency of Mines and gave courses in mineralogy and physics at *L'Ecole des Mines* from 1795 until 1802. During this period he was also a curator of the mineralogical collection, a position of invaluable assistance to him in the completion of his *Traité de minéralogie* because of the variety and extent of the mineralogical specimens that it contained. When the Institut National des Sciences et des Arts was organized to replace the former academies in 1795, Haüy was named one of the first two academicians in the section of natural history and mineralogy. In 1802 Haüy was appointed to the chair of mineralogy at the Museum of Natural History, where he remained until his death.

After the publication of the *Traité de minéralogie*, Haüy gained scientific fame throughout Europe. He carried on an active correspondence with contemporary scientists and became a member of more than twenty foreign scientific societies and academies. When the Concordat of 1801 was ratified, Napoleon named Haüy honorary canon of the cathedral of Notre Dame, and in 1803 the first consul extended Haüy membership in the Legion of Honor as a reward for writing the *Traité de physique*, designed as a text for all national *lycées*. Haüy died on June 1, 1822, after sustaining a broken femur.

Haüy's students and followers multiplied his researches. Some of the most noted were Alexander Brongniart (1770-1847), Louis Cordier (1777-1861), François Sulpice Beudant (1787-1850), and Gabriel Delafosse (1796-1878). Delafosse, in turn, was the teacher of Louis Pasteur (1822-1895). In crystallography Pasteur made the discovery, so important for stereochemistry, of the enantiomorphism of the crystals of sodium ammonium tartrate and racemate by linking their observed hemihedry with their optical properties.[12] Haüy's theory of crystal structure was taught in French schools relatively soon after his initial publications, and contemporary writings emphasize the impetus his work gave to the study of mineralogy and to an interest in it on the part of amateurs.[13]

His biographers confessed, however, that Haüy had one failing. He could not accept objections to or criticisms of his theory with good grace or scientific detachment. Upon such occasions even his sleep became troubled, and his usual calm demeanor was lost. In this respect, Haüy's attitude provides a good example of the fallacy of the popular image of the scientist as not being emotionally in-

---

[12] For a good account of this phase of Pasteur's work see J. D. Bernal, *Science and Industry in the Nineteenth Century* (London, 1953), pp. 183-202.

[13] For example, it was taught at the *école centrale* at Lyon in 1796. See Louis Trénard, *Lyon* (2 vols.; Paris, 1958), II, 490. For its effect on the interest in mineralogy see *Dictionnaire des sciences naturelles*, ed. F. G. Levrault (Paris, 1818), XI, 432. Élie Halévy, *England in 1815* (London, 1949), p. 564, documents the following information concerning William Allen: "About this period (1804) W. Allen attended a series of conversazione at Dr. Babington's, where Count Bournon gave instructions in mineralogy, particularly crystallography."

volved in his work. In addition, these biographical comments illuminate the tone that characterizes Haüy's writings. He was confident that his theory and conclusions were correct to the last degree, and, to the end of his life, he refused to admit the legitimacy of criticisms he could not scientifically contradict. In part, his conduct may be attributed to the ordinary human aversion to suffering criticism and admitting error, but, in a large measure, it was due to his rigid preconceptions of the nature of the structure of crystalline matter and of the importance of crystal form in the delineation of a mineral species. In order to elucidate these remarks, it is necessary to present Haüy's structural theory.

To account for the variety of secondary crystal forms which any substance might manifest, it was first necessary for Haüy to determine the shape of the basic building block of the substance and the values of its plane and interfacial angles.[14] Normally, careful cleavage of a secondary form or crystal variety was performed to a point where it appeared that the resulting configuration would suffer no change by additional cleavage. In the early expositions of his system, Haüy stated that this resulting crystal was macroscopically equivalent to the constituent molecules (*molécules constituantes*) of the crystal. The constituent molecules were identical to one another and to the nucleus of the crystal. If the constituent molecule was a parallelepipedon, as was true, for example, with calcite, Haüy had no difficulty in using it to construct the crystal varieties. But as time went on, Haüy found there was a variety of nuclei; he therefore discarded the term "constituent molecule." Cleavage of a variety of substances showed Haüy that the number of different nuclei could be reduced to just six diverse shapes which he called primitive forms. These were the parallelepipedon, the dodecahedron with rhombic surfaces, the dodecahedron with isosceles triangular surfaces, the right hexagonal prism, the octahedron, and the tetrahedron. These shapes, he asserted, were the true primitive forms discovered by mechanical division of any crystal and not the postulated primitive forms of Romé de l'Isle. Although the nucleus was the last term in the mechanical division of any crystal, it was capable of mathematical subdivision; thus the six primitive forms could be reduced in

---

[14] The following description is drawn from Haüy, *Essai*; Haüy, "Exposition de la théorie sur la structure des cristaux," pp. 225-319; and Haüy, *TDM1*, I, 19-109; II, 249-255.

number to just three configurations or integrant molecules (*molécules intégrantes*). For example, in the instance of garnet whose primitive form was a dodecahedron with rhombic surfaces, Haüy supposed at first that it could be considered to have been formed from four congruent rhombohedra, each of which had three of its faces touching a similar face of the other three. But each of these rhombohedra could, in turn, be constructed by the juxtapositions of six tetrahedra, so that the tetrahedron became the integrant molecule of garnet. The dodecahedron with isosceles triangular surfaces could be constructed similarly from tetrahedra, and the right hexagonal prism from triangular prisms. Thus, the six primitive forms were reduced to four: the parallelepipedon, the tetrahedron, the triangular prism, and the octahedron. The octahedron presented a difficulty because it is not possible to build it up from tetrahedra without leaving voids. There is, however, a relationship between the octahedron and the tetrahedron; for example, a tetrahedron may be considered as being derived from an octahedron by juxtaposing four tetrahedra on alternate faces of an octahedron. Haüy decided to discard the octahedron as an integrant molecule, thus reducing the number of primitive forms to three: the tetrahedron, the triangular prism, and the parallelepipedon. In making this decision, he was impressed by the fact that these configurations had the greatest degree of simplicity in that they had four, five, and six faces, respectively.

Yet in order to be able to work with parallelepipeda exclusively as the building blocks of the crystal structure, and to avoid the difficulties encountered with the octahedral primitive form, Haüy postulated another type of molecule. The juxtaposition of two triangular prisms will produce a parallelepipedon, a rhombohedron can be constructed from six tetrahedra, and a combination of an octahedron and two tetrahedra will form a rhombohedron as well. Haüy called these fictional units subtractive molecules (*molécules soustractives*). By positing the subtractive molecules, Haüy believed that his theory received a higher degree of generality because they performed the same function as the usual parallelepipedal integrant molecules produced by ordinary cleavage. They, too, became unit building blocks of the crystal structure.

After determining the shape of the primitive form and the integrant molecule, it was necessary to assign values to the plane and interfacial angles of the latter. The angles of inclination of the cleavage faces to the crystal axis were determined, and with these values, Haüy, using plane trigonometry, could calculate the ratios of the diagonals of each face and, finally, the values of the facial and interfacial angles. Haüy's treatment of the calcite integrant molecule that was identical with the Iceland crystal is a good and important example of this procedure.

Cleavage of that secondary form of calcite which consists of a hexagonal prism combined with the basal pinacoid (faces perpendicular to the axis of the prism) seemed to indicate to Haüy that the cleavage faces of the resulting rhombohedron were inclined at *exactly* 45 degrees to the axis of the crystal. From this value he was able to calculate the ratio of the diagonals of the rhombic faces as $\sqrt{3}:\sqrt{2}$. In turn, the calculated value of the larger angle of the rhombic face was 101° 32' 20" and that of the interfacial angle was 104° 28' 40".[15] (See Appendix I for the complete calculation.)

It is evident that all Haüy's calculations rested on the assumption that the angle of inclination of the cleavage face to the crystal axis was exactly 45°. Haüy did not point out this fact, but his contemporaries recognized that his procedure, in fact, assigned a value to the angle the face made with the axis. Daubenton and Laplace recognized this assumption in their report to the Academy of Sciences on the content and value of Haüy's *Essai d'une théorie sur la structure des crystaux*. But they remarked that this supposition, in the case of calcite, led to a value for the obtuse angle of the plane faces which differed by only 2' 13" from the value of 101° 30' *which de la Hire had assigned to it in 1710*. The report concluded: "It is certainly probable that this slight difference stems from an error in [de la Hire's] observation, and that the result of Abbé Haüy coincides exactly with that of nature."[16] On this point, however, Romé de l'Isle took issue with the "cristalloclaste," as he contemptuously termed Haüy. According to the former, it would be impossible for anyone to infer that the rhombohedra that formed a hexagonal prism of calcite could be precisely the same figure as the Iceland crystal. It seemed obvious to him from the consideration of a section through a regular hexagonal prism that such a shape could be produced only by the

---

[15] Haüy, *Essai*, pp. 95-97. Peculiarly, Haüy does not even refer to the law of the constancy of interfacial angles in this work. Romé de l'Isle had stated the law in 1783, just a year earlier.

[16] L. Daubenton and P. Laplace, "Rapport de l'Académie des sciences sur l'essai d'une théorie sur la structure des crystaux," *JDP*, XXIV (1784), 71-74.

juxtaposition of three rhombohedra with angles of 60° and 120°. Romé de l'Isle did not seem able to follow Haüy's demonstration. Cleavage and calculation, in his opinion, could not take the place of direct observation and goniometric measurements of the crystal. As far as he was concerned, the value of the obtuse facial angle of the Iceland crystal was 102° 30', Professor Haüy to the contrary notwithstanding. Including the Academy of Sciences in his castigation, Romé de l'Isle concluded:

> This example ought to suffice to put us on guard against these pretended geometrical demonstrations over which there is so much uproar, because they have received the approval of an academy whose most respected members are also too enlightened to believe that they have received universal knowledge.[17]

Romé de l'Isle's objections and comments not only showed a complete misunderstanding of Haüy's method and lack of sympathy for his purpose, but also failed to indicate the basic assumption Haüy made. Since the values of the facial and interfacial angles which Haüy calculated for the varieties of calcite and other crystalline substances were in general agreement with the measurements of the contact goniometer on actual crystals, Haüy could shrug off such diatribes easily and proceed confidently with the extension of his theory. Also, his work appeared more authoritative because, using six- or eight-place logarithms, he reported all angular values to the nearest second or half-second, whereas the accuracy of the contact goniometer was no better than one-quarter of a degree.

By 1788, however, when he wrote a memoir on the double refraction of calcite, Haüy was aware that Huygens had listed different angular values for the Iceland crystal, specifying particularly that the inclination of the faces to the axis was not exactly 45°. But he stated confidently: ". . . this savant [Huygens] appears to me to have increased somewhat the measurements of the principal angles of the rhombohedron. . . . I believe that I have succeeded in making more precise measurements."[18] Somewhat later it was more difficult for Haüy to discount the meticulous observations of William Hyde Wollaston (1766-1828), who reported

---

17 J. B. L. Romé de l'Isle, *Cristallographie, ou description des formes propres à tous les corps du règne minéral* (4 vols.; Paris, 1783), I, xxx, 494.

18 R. J. Haüy, "Mémoire sur la double réfraction du spath d'Islande," *ADS* (1788), p. 42.

in 1802 that careful measurements indicated the interfacial angle of the Iceland crystal had a value of 105° 5'.[19] (See accompanying table.) From this value Wollaston calculated that the obtuse angles of the rhombic faces were 101° 55', and that the angle the

REPORTED VALUES OF THE PRINCIPAL ANGLES OF THE ICELAND CRYSTAL

Source	Obtuse facial angle	Interfacial angle	Inclination of faces to vertical axis
Huygens	101° 52'	105°	45° 20'
De la Hire	101° 30'	105°	
Romé de l'Isle	102° 30'		
Haüy	101° 32' 13"	104° 28' 40"	45°
Wollaston	101° 55'	105° 5'	45° 23'
Modern value	101° 55'	105° 5.07'[a]	45° 22.81'

[a] The modern notation of the fundamental interfacial angle (rr') is the supplement of that listed, i.e., 74° 54.93'. Dana's *System of Mineralogy* (7th ed.; New York, 1944), II, 158, reports the best modern values as those given by Charles S. Hastings, "On the Law of Double Refraction in Iceland Spar," *American Journal of Science*, 135 (1888), 60-73.

surfaces made with the axis was 45° 23'. When Wollaston perfected the reflecting goniometer in 1809, he offered almost incontrovertible proof that the interfacial angle was nearly, if not exactly, 105° 5'.[20] This result was confirmed within a short time by Etienne L. Malus (1775-1812), whose studies of the double refraction of calcite caused him to perfect a crystal-measuring device similar to Wollaston's.[21]

Haüy remained unconvinced by these observations. First, he believed that the measurements given by the reflecting goniometer were derived indirectly; in other words, the instrument itself was a mechanical device subject to error, however slight. Second, all crystals were susceptible to minute accidental deviations in shape which occasioned much greater differences in angular values than his critics had noted in the Iceland crystal. But

[19] W. H. Wollaston, "On the Oblique Refraction of Iceland Crystal," *PT* (1802), pp. 381-386.
[20] W. H. Wollaston, "The Description of a Reflective Goniometer," *PT* (1809), pp. 253-258.
[21] E. L. Malus, *Théorie de la double réfraction de la lumière dans les substances cristallisées* (Paris, 1810), pp. 99-101.

the most important factor in Haüy's decision to ignore these observations was that their acceptance would destroy a simple mathematical relationship which, to Haüy, had the character of a limit. If his value of the interfacial angle was used, the ratio of the diagonals of the rhombic faces of the calcite integrant molecule was $\sqrt{3}:\sqrt{2}$. On the other hand, the simplest ratio of these diagonals which could be calculated by using the value of 105° 5' for the interfacial angle was $\sqrt{111}:\sqrt{73}$. Simplicity, Haüy asserted, was demanded by science, and therefore it was necessary to abandon such observations as a sacrifice commanded by science. In addition, Haüy asked, what differences were occasioned in other varieties of calcite by the insistence on an integrant molecule of a slightly diverse shape? In the "inverse" rhombohedron, the value of the interfacial angle would become 101° 8' 56" in place of his value of 101° 32' 13", a difference of only 23 ' 17". In the dodecahedral scalenohedron, the incidence of the triangular faces would become 144° 24' 16" instead of 144° 20' 26", a difference of only 3' 50". According to Haüy, these disparities were insensible; they could and should be neglected because his calculations were based on simple relationships.[22] To Haüy, nature had formed crystalline matter in accordance with the principles of simple arithmetic and geometry; we should not attempt to complicate matters once these simple relationships had been established.

Haüy also used this concept of simple limit ratios to derive the value of the interfacial angles, and, consequently, the shape of the integrant molecules of those substances in which cleavage was absent or practically so. For example, in the case of quartz (fig. 9), he measured with a contact goniometer the inclination of one of the faces of the rhombohedron, such as P, on the adjacent face r of the prism. He found that the angle was between 141° 30' and 142°, and he supposed that it was 141° 45'. If, from the center of the base c, the line cr is drawn perpendicular to the side, and r is joined to s, the angle crs would then be equal to 51° 45'. Since the triangle crs is a right triangle, $cr:cs :: \sin 38° 15' : \sin 51° 45'$. Taking the square roots of the squares of these sines, Haüy obtained the ratio, $cr:cs :: \sqrt{3833}:\sqrt{6167}$. He considered that this ratio was nearly $\sqrt{38}:\sqrt{62}$ or $\sqrt{19}:\sqrt{31}$, and, if unity was added to each term under the square root signs, the ratio became $\sqrt{5}:\sqrt{8}$.

[22] Haüy, TDC, II, 386-395. Also see Dictionnaire des sciences naturelles, XI, 565n.

In Haüy's opinion, the latter ratio had the requisite simplicity to give it the character of a limit. Using this ratio Haüy calculated the value of the interfacial angle of P and r to be 141° 40' 16", from which the measured value deviated 4' 44". This value, then, approached that obtained by mechanical measurement, and, because of the simplicity of the ratio from which it was derived, Haüy believed that this value was that which the processes of nature attempted to duplicate in any crystallization of quartz.[23]

Fig. 9. Haüy's sketch of quartz. SOURCE: *R. J. Haüy, "Observations sur la mesure des angles des cristaux," JDP, LXXXVII (1818), facing p. 253.*

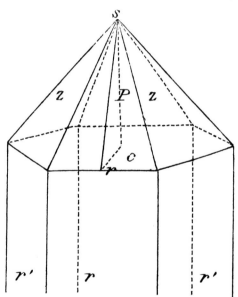

Haüy attempted to justify his procedure in a number of ways. He stated that if he showed a mathematician the ratio $\sqrt{149}:\sqrt{240}$ and informed him that it was derived from a mathematical consideration of the mutual inclination of two faces of quartz, it would be very probable that the mathematician would make a small correction that would allow both terms under the square

[23] R. J. Haüy, "Observations sur la mesure des angles des cristaux, *JDP*, LXXXVII (1818), 233-253.

root sign to be divided by 30, hence reducing the ratio to $\sqrt{5}:\sqrt{8}$. If the mathematician knew that the difference in inclination resulting from his operation amounted to only 4', Haüy did not doubt that he would be inclined to attribute the discrepancy to observation rather than to nature. Haüy referred to the experiments of Charles A. de Coulomb (1736-1806), wherein the electrical charges of bodies and the distances between them were related. Haüy explained that Coulomb's experiments had never rigorously demonstrated the operation of the law of inverse squares between the charges and the distance, but they approached the law so closely that Coulomb was authorized to use it. In Haüy's opinion the progress of science necessitated the synthesis of theory and observation in order to eliminate the lack of connection which would otherwise exist between the results of the two approaches. The measurements of angles with the reflecting goniometer, he said, were isolated observations only. No one had attempted to bring them together into a requisite form suitable for the construction of a theory.

By stipulating that the dimensions of the integrant and subtractive molecules must be mathematically simple, Haüy introduced errors into his theory which not only caused embarrassment but also contributed to its eventual replacement. It is probable that his insistence on this feature was caused by the apparent mathematical simplicity of the laws according to which the integrant molecules were superimposed upon one another. Haüy named these relationships the laws of decrement (*décroissement*), each of which, either individually or in combination with one another, served to explain the appearance of any crystal form in any substance. Every crystal face could be considered as having been built up by the progressive addition of lamellae to it. Each lamella was thought to be formed of contiguous integrant or subtractive molecules and to have the thickness of one molecule. Haüy's first law of decrement involved the successive subtraction of integrant molecules from the edges of each newly added layer. Assuming that the nucleus of a crystal is a cube and that the added integrant molecules are also cubes, if one row or rank of cubic molecules is subtracted from each successively applied layer, the cube will eventually take the form of a dodecahedron with rhombic faces. Thus, if the law of decrement by one row of the edges began to operate on a cube composed of 729 molecules—81 on each face of the cube—the first layer applied to each face would contain only 49 cubic molecules, the second 25 molecules, the third 9 molecules, and the final layer one molecule (fig. 10). Haüy explained that the minuteness of the molecules precluded even microscopic observation of the stairlike aspect of the successively smaller lamellae. The growing faces presented an even external appearance. If the growth of the crystal ceased before the dodecahedron was completed, the crystal would have a cubic shape, but, in place of edges, there would be elongated hexagons inclined at 45° angles to the cubic faces.[24]

Fig. 10. Haüy's illustration of how the dodecahedron with rhombic faces is constructed by the progressive decrement of one row of molecules on each edge of lamellae successively added to a cubic nucleus. SOURCE: R. J. Haüy, Traité de minéralogie (5 vols.; Paris, 1801), V, Pl. II, figs. 11, 13.

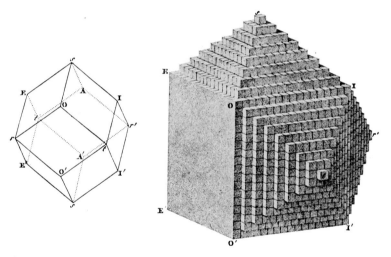

In many instances Haüy was required to postulate differential decrement on the edges. For example, there might be a decrement of two rows of cubic molecules on two opposite edges of each successively deposited layer, while a decrement of only one row of molecules on every other new layer occurred on the other two edges. When decrement operated in this manner with cubic in-

_____

[24] Haüy, "Exposition de la théorie sur la structure des cristaux," pp. 237-243.

*Fig. 11. Haüy's illustration of how the
dodecahedron with pentagonal faces is constructed
from a cubic nucleus by the combined
decrement of two rows of molecules in width
between the edges OI and AE, II'
and OO', EO and E'O', and simultaneously by two
rows of molecules in height between
the edges EO and AI, OI and O'I', OO' and EE'.*
SOURCE: *R. J. Haüy,* Traité de minéralogie
(5 vols.; Paris, 1801), V, Pl. II, figs. 14, 16.

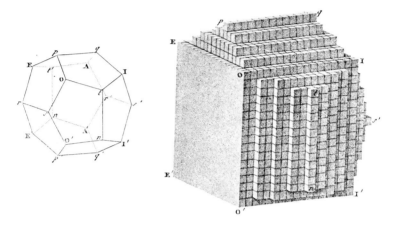

tegrant molecules, the resulting crystal was a dodecahedron with pentagonal faces, the form of pyrite crystals (figs. 11, 12). He did not state why differential decrement of this type occurred. He was interested only in explaining how a particular crystal shape could be produced by the superimposition of lamellae composed of integrant molecules on a primitive form or nucleus. Because the molecules were units, the plane angles and the ratio of the sides of the faces, as well as the interfacial angles of the resulting crystal, could be calculated by the methods of plane trigonometry. He demonstrated that the regular dodecahedron and icosahedron of classical geometry were impossible crystallographic figures because the decrement ratios calculated from the interfacial angles of these figures were incommensurable numbers. These forms, then, could not be produced in nature because the decrements operated by definite rows of molecules.

The second general law of decrement operated on the angles or parallel to the diagonals of the primitive form. Haüy explained he had derived this law by observing that the same substance that

Fig. 12. Haüy's illustration showing a kind of
pyramid which is the effect of decrement by two
rows of molecules on lamellae added
to a rhombohedral nucleus. SOURCE: R. J. Haüy,
Traité de minéralogie
(5 vols.; Paris, 1801), V, Pl. III, fig. 17.

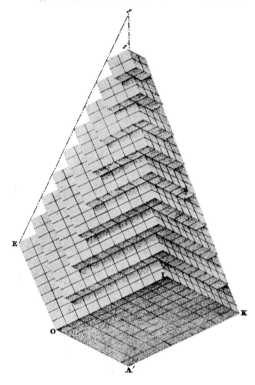

crystallized as dodecahedra with pentagonal or rhombic surfaces
and originated from a cubic nucleus also crystallized as octahedra.
He had believed at first that it would be possible to derive the
structure of the octahedron by limiting the decrements to the
edges of two opposite faces of the cube. Two pyramids would be
applied on these faces, and, if the pyramidal faces were prolonged
until they met in apexes, an octahedron would be produced. This
construction demonstrated to Haüy that there was no law, however complicated, by which equilateral triangles could be generated on the octahedron. If a regular octahedron were analyzed
structurally, according to Haüy, one could observe that the cubic
nucleus was situated within the octahedron in such a way that

each of the eight solid angles of the cube corresponded to the center of one of the faces of the octahedron. The fact caused him to develop a new group of laws.

In order to construct the regular octahedron from the cube, Haüy posited that it was necessary to apply the law of decrement on all solid angles of the cube. Assuming again a cubic nucleus composed of 729 cubic molecules so that each face showed 81 molecules, as the law commenced to operate a row of molecules was added to each edge of the first two superimposed lamellae while at the same time molecules were subtracted from the corners of the layers along lines parallel to the diagonals. The second layer was thus in the shape of a diamond, reaching a maximum length and width of 13 molecules. Thereafter, each successive superimposed layer decreased by one row of molecules parallel to the diagonals at each of the original corner angles. The application of such lamellae would result in the construction of the required octahedron, but Haüy obviously had to account for the occurrence of the first two lamellae. He admitted that this supposition was required because otherwise reentrant angles would be formed at the edges of the cube. He explained that possibly the crystal grew in those parts to which the laws of decrement did not extend, but he emphasized that he did not believe crystal growth took place exactly in the required manner. He admitted he was describing the artificial construction of a solid that represented a crystal and felt that such abstraction must be made in order to account for all details of a crystal structure (see Appendix II).[25]

Decrement of molecules at the edges and angles of a nucleus corresponding to the basic shape of the integrant or subtractive molecules were Haüy's two fundamental laws. But there could be, in addition, intermediate decrements. In certain crystals, decrement at the angles took place parallel to lines situated between the diagonals and the edges rather than parallel to the diagonals. In other crystals, decrements either at the edges or at the angles could be mixed. It could happen that the superimposed lamellae decreased by two rows of molecules at the edges and were, at the same time, triple the thickness of a simple molecule. Sometimes the decrements operated at certain sides and angles and not at others. In rare instances the same edge or angle was affected by several laws of

25 *Ibid.*, pp. 250-268.

decrement which operated successively. Haüy noted that the forces causing the observed decrements seemed to operate in a very limited manner. Subtractions were almost always confined to one or two rows of molecules. Decrement entailed no more than four rows except in one variety of calcite where the subtraction extended to six rows. If the laws of decrement were limited to one or two rows of molecules, Haüy calculated that 2,044 various forms of calcite were possible. If the subtraction of three and four rows of molecules was admitted, however, the laws would allow 8,388,604 possible forms of this substance.[26]

Haüy's laws of decrement worked in practice because the observed faces on dominant crystal forms are usually parallel to planes having high reticular densities, that is, planes through many nodes of a crystal lattice, with Haüy's integrant molecules viewed as being at the nodes. Not visualizing the crystal lattice, however, Haüy could not, of course, know that there is a much greater probability that certain faces, because of the higher reticular densities, will be formed and that the probability of the formation of many other faces is quite low. His calculation as to the possible number of calcite varieties reflects his sentiment that all possible faces might be formed with equal probability.

Haüy devised a shorthand system to delineate any face of a secondary crystal form. In it, plane faces, angles, and edges of the primitive forms were designated by the capital letters of the alphabet. Vowels were used to denote the solid angles, and consonants were used for the edges. Angles and edges higher in a sketch of the primitive form took precedence in the placement of the letters with the designation proceeding from left to right (fig. 13). Lowercase alphabetical letters were used for the symmetrically positioned angles and edges of the lower half of the form. Accompanying the letters were numbers that indicated the law of decrement which operated at the angles or edges to produce a secondary face. Decrements of one, two, or three rows of molecules in length or width were indicated by the integers 1, 2, or 3. Decrements of one, two, or three rows of molecules in height were expressed by the fractions $1/1$, $1/2$, and $1/3$. For example, O^2 denoted the effect of a decrement by two rows of molecules parallel to the diagonal of the base P at the angle O. O^3 represented the effect of a decrement by three rows of molecules parallel to

Fig. 13. The basis for Haüy's system of nomenclature. SOURCE: *R. J. Haüy*, Traité de minéralogie *(5 vols.; Paris, 1801), V, Pl. VI, fig. 48.*

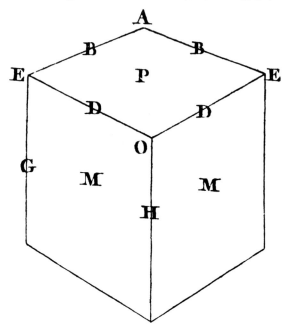

the diagonal of face T which passed through the angle I. O^4 showed a decrement of four rows of molecules parallel to the diagonal of face M which passed through the angle E. $\overset{2}{D}D$ denoted the effect of two simultaneous decrements on the edge D, one by two rows of molecules on the face M. The use of parentheses demonstrated the presence of intermediate decrement. For example, $(O^1D^1F^2)$ indicated that decrement took place by one row of molecules on the angle O on the face $AEOI$, and that for one row of molecules subtracted along the edge D, two rows were along the edge F. As an example of a complete form, a calcite scalenohedron was indicated by the combination $(E^{1\ 1}ED^1D^2)$ $\overset{2}{D}E^{1\ 1}E$. The notation was cumbersome but workable.[27]

The first basic rule in Haüy's method was that the choice of the structure or the primitive form of the integrant molecule should not be arbitrary as it was with Romé de l'Isle. If the structure could not be ascertained definitively by mechanical division, its

[27] Haüy, *TDC*, I, 257-280.

postulation was permissible. There must be, however, some reasonable basis for speculation; striations on the faces, for example, might be the basis for an educated guess. There was a problem, however, when a substance cleaved in more than three directions. Fluorspar, for example, has perfect octahedral cleavage, so that Haüy's final division yielded an octahedron, and it was to take care of such phenomena that Haüy conceived of the subtractive molecule. But even when the integrant molecule could be obtained by mechanical cleavage, there was the further precept that it should possess mathematical simplicity. Cubic integrant molecules, of course, enjoyed this property in the highest degree. Rhombohedral molecules were expected to conform in the same way, it being mandatory that the ratio of their diagonals be expressed in small numbers that gave the appearance of a limit. When this relation was obtained, the basic structure of the integrant molecule could be indicated by denoting the values of the angles of the faces and the interfacial angles.

Haüy treated the crystallization of the integrant molecules into geometric solids in a completely static manner. He stated that his purpose was not to explain the metamorphoses of crystals by reference to the conditions of crystallization.[28] He supposed that the integrant molecules reunited by virtue of the attractive force that acted between them, but he was not interested in speculating on the nature of this force. Crystallization was not considered in the context of a dynamic process; instead, it was merely the capability of identical integrant molecules to give birth to a multitude of different forms. The purpose of the study of crystallization, then, was to determine the mathematical relationships that existed between primary and secondary forms. There was a connection between a face of a secondary crystal of each species and a part of the primitive form. The work of the crystallographer entailed the determination of the laws that linked these corresponding parts. More exactly, the inclination of a secondary crystal face to a corresponding face, edge, or solid angle of the primitive form depended upon a law that it was the task of the crystallographer to uncover and state. A second fundamental rule of Haüy's method stated that these laws of linkage must be demonstrated with mathematical rigor. What was the nature of these laws? First, Haüy's method included the concept that the integrant or subtractive

---

[28] Haüy, *Essai*, p. 8.

molecule possessed a determinate shape. The ratio of the diagonals of this configuration and, consequently, the values of the facial and interfacial angles were known. Second, the laws of decrement which explained the passage of an aspect of the primitive form to the face of a secondary form operated by the subtraction of single or multiple rows of whole integrant or subtractive molecules. If the goniometer measurements of the plane and interfacial angles of a secondary form agreed exactly with the angular values calculated from the application of the particular law or group of laws to the primitive form, there was proof that the correct laws of decrement had been used and that the conceived form of the integrant molecule was an exact representation of nature. If the observed and calculated angles did not agree, two possibilities were open to investigation: an incorrect method or law had been employed to derive the secondary form, or the true integrant molecule was not the one that had been postulated. If the trial of many laws of decrement did not establish the required face of the secondary form with any degree of accuracy, then the integrant molecule required additional investigation. The experimental part of Haüy's theory, then, was the establishment of the shapes of the integrant molecule and the primitive form and the application of the correct laws of decrement in order to develop the secondary form. Under no circumstance could the assumption of decrement by rows of fractional molecules be admitted. The laws of decrement were a priori, then, to the extent that they had to operate by rows of whole molecules only. This concept sprang directly from the fundamental assumption that the planes of cleavage were the layers of application of integrant molecules. Additional indication that the correct laws had been used could be elicited from mechanical division and observation of striations on the faces of the secondary forms.

The period during which Haüy developed his theory of crystal structure and extended his observations to include hundreds of mineral species was contemporaneous with the scientific upheaval in chemistry and with the resurgence of interest in physical optics. Research in both disciplines produced results that necessitated major revisions in Haüy's theory. These studies are of such importance that they will be treated in detail in later chapters. Apart from these major difficulties, however, Haüy's theory and method were questioned in other respects. His dictum that the dimensional ratios of the integrant molecules must be mathematically simple, that they must possess the character of a limit, came more and more into question as the angular values of crystals were observed and measured with the reflecting goniometer. As in the case of the calcite rhombohedron, these experimental results did not shake Haüy's confidence in the truth of his concept. But they did provoke serious doubts concerning the foundations of his theory in the minds of others. Wollaston claimed that the "apparent harmony by simple ratios was a seducing circumstance."[29] The English crystallographer, H. J. Brooke (1771-1857), stated in 1819:

His [Haüy's] theory contains a principle which has in some instances conduced to error and which may affect its worth as a theory more than any consideration of the comparative merit of the goniometer. This principle is an imaginary simplicity which he supposes to exist naturally in the ratios of certain lines either upon or traversing a crystal ... and he is disposed to regard generally the disagreement of an observed measurement with this character rather as an error of the observation than a correction of his theoretic determination.

Brooke concluded: "We may hope that the second edition of his *Mineralogy* . . . will rather adopt and reason upon the dimensions that nature has given to crystals than clothe them with an imaginary character."[30] Impatience on this point reached a peak in the statement of Frederick Mohs (1773-1839) in 1823: ". . . the measurements, or rather the indications of the angles of Haüy, have, in so many instances, been found incorrect, that we can no longer attach any certainty to their exactness."[31]

In some instances, Haüy's theory became quite arbitrary. This defect was even acknowledged in the glowing report of the progress of science in France which the Institute was ordered to prepare for Napoleon. One instance, for example, arose when the primitive form was a hexagonal prism. Mechanical division could not, under any circumstances, give the ratio of the height of the prism to the width of its base. It was necessary to choose a suitable ratio and then determine whether the secondary forms developed by the laws of decrement corresponded to nature.[32]

---

[29] Wollaston, "Description of a Reflective Goniometer," p. 256.

[30] H. J. Brooke, "Observations on a Memoir by the Abbé Haüy on the Measurement of the Angles of Crystals," *AP*, XIV (1819), 453.

[31] F. Mohs, "On the Crystallographic Discoveries and Systems of Mohs and Weiss," *EPJ*, VIII (1823), 289.

[32] G. Cuvier, ed., *Rapport Historique sur le progrès des sciences naturelles depuis 1789 et sur leur état actuel* (Paris, 1810), p. 17.

In addition, despite Haüy's emphasis upon structure as a geometrical arrangement of integrant molecules, the theory had
physical implications that disturbed other scientists. For example,
Haüy's construction of the octahedron from the cubic nucleus
supposed that the molecules were added at the edges of the initial
lamellae juxtaposed on the faces of the cube and that molecules
were being subtracted simultaneously at the corners or angles of
the same lamellae. Haüy explained that this hypothesis was necessary in order to prevent the formation of reentrant angles on the
faces of the growing octahedron. But this type of growth did not
occur in calcite, for example, when it was being transformed from
the rhombohedron into the hexagonal prism. In essence, Haüy pre-
supposed the existence of secondary surfaces which his theory was
designed to prove. He attempted to counter this defect by repeated
statements that the formation of the crystal did not take place in
the manner indicated by the laws of decrement. He stated he was
far from thinking that an octahedron of sulfur or quartz commenced by being a cube of a volume proportional to that of the
octahedron and then grew by the addition of successive lamellae,
passing through all the stages between the cube and the octahedron
as he indicated in his sketches. On the contrary, he said he believed
that the octahedron was formed by the growth of an imperceptible
octahedron composed of the lowest possible number of molecules.
The tiny octahedron, even at this stage, had a cube as a nucleus.
As the octahedron grew by the addition of new lamellae of cubic
molecules superimposed on its faces, the central cube also grew.
In this manner, the geometric structure of the crystal remained
the same.[33] Thus, one contemporary mineralogical text reluctantly
concluded: "We cannot indeed demonstrate that secondary forms
are actually produced in the manner, which the theory supposes.
It is however no inconsiderable argument in its favor, that all calculations, founded on it, give results perfectly conformable to
observed facts."[34]

For several reasons, the notion of the subtractive molecule was
particularly disconcerting. First, there was the uncertainty of the
structure of many substances. According to Haüy, some subtractive molecules could be considered as being composed either of

---

[33] Haüy, *Essai*, pp. 206-219; R. J. Haüy, "Exposition abregée de la théorie
de la structure des cristaux," *ADC*, III (1789) 16-17.
[34] Parker Cleaveland, *An Elementary Treatise on Mineralogy and Geology*
(Boston, 1822), p. 21.

tetrahedral molecules with octahedral voids or of octahedral molecules with tetrahedral voids. Second, why must the presence of interstices between the integrant molecules of crystals having octahedral cleavage be accepted when this condition was not presumed in the structure of crystals having cubes or rhombohedra as integrant molecules? No explanation was forthcoming from Haüy. The concept of the subtractive molecule was designed to have a definite geometrical meaning only, and no physical significance should be attached to it. But even Haüy's adherents could not abstain from criticizing his intransigence in giving a physical clarification of this notion. The supposition of voids was gratuitous, and this defect in Haüy's theory had to be admitted.[35]

It is important to note also that Haüy believed his theory of crystalline matter was complete. It was necessary, of course, that a great deal more work be done. There were hundreds of crystalline substances whose structures remained to be investigated. But, by following his precepts and his method, these unknown structures could eventually be identified. This point of view was not shared by all scientists. One review, though admitting that Haüy's system furnished excellent principles, pointed out that the real theory of crystal structure could not be considered understood until the law of force by which the regular arrangement of molecules was brought about was known, as well as the cause that determined the rate of decrement of the lamellae.[36] Later, when reporting the results of research into the primitive form and crystal modifications of tin oxide, a reviewer in the same journal stated:

We do not mean to question the utility of minute crystallography, as a part of the science; yet, we own that it is somewhat painful to us to see labour and ingenuity so largely vested in this research; and it is impossible not to view most of its results as mere *culs-de-sac*, out of which we return without either profit or pleasure.[37]

It was admitted universally that Haüy had succeeded in explaining the reason for the occurrence of the variety of forms which the crystals of the same substance might take by the assump-

---

[35] Cuvier, *loc. cit.*; *Dictionnaire des sciences naturelles*, XI, 540.

[36] Anonymous, "Discours sur le progrès des sciences, lettres, et arts depuis 1789, jusqu'à ce jour (1808); ou Compte rendu par l'Institut de France à S. M. l'Empereur et Roi," *Edinburgh Review*, XV (1809), 16.

[37] Anonymous, "Transactions of the Geological Society," *Edinburgh Review*, XXVIII (1817), 188.

tion of integrant molecules and laws of decrement which affected their superimposition on the crystal nucleus. The genius that had created this valuable scientific theory from so small a number of material facts was appreciated fully. Haüy's calm scientific approach to the study of crystal structure was in complete accord with the contemporaneous scientific method. But there were, in addition, the misgivings that have been noted concerning his theory, and many scientists felt that Haüy's theory lacked the generality and the completeness that he attributed to it. These qualms existed twenty years before Haüy's death and were held by some who had received their crystallographic training from him. The results obtained by probing the weak spots in Haüy's theory and by questioning his static treatment of the structure of crystals had important consequences for the future not only of the science of crystallography but also of chemistry.

CHAPTER V

Crystals in Chemistry and Optics

The discovery of isomorphism and polymorphism and the recognition of a definite relationship between optical properties and crystal form were among the principal causes for the abandonment of Haüy's crystallographic method and the mineralogical system he had based upon it. The work of Nicolas Leblanc (1742-1806), Johann Fuchs (1774-1856), William Hyde Wollaston, and François Sulpice Beudant (1787-1850) progressively produced evidence of the existence of what are now termed solid solutions, wherein the ions of an element may be substituted in a chemical compound for ions of another element. One can perceive also in some of this research the embryonic form of the idea of isomorphism, that chemically related substances show a close similarity of crystal form. The existence of this phenomenon was first stated clearly by Eilhard Mitscherlich (1794-1863). It was Mitscherlich, also, who made the first thorough study of polymorphism, the phenomenon wherein a chemical substance crystallizes into two or more distinct atomic arrangements, although this had been previously recognized indistinctly by a number of scientists. Similarly, Johann Jacob Bernhardi (1774-1850) and Etienne Malus prepared the foundations for research on the optical properties of crystals, the former by stressing the directional aspect of the phenomenon of double refraction and the latter by discovering the polarization of light. The harvest was reaped, however, by Jean Baptiste Biot (1774-

1862) and David Brewster (1781-1868), who independently discovered optical biaxiality and employed this property in the analysis of crystalline substances. The present chapter considers the work of these men in the areas of chemistry and optics and delineates how the results of their efforts affected the science of crystallography as it had been erected by Haüy.

Implicit in Haüy's theory of the structure of crystals as consisting of the juxtaposition of identically shaped molecules was the idea that the molecules were also qualitatively or chemically identical. In his 1784 *Essai d'une théorie sur la structure des crystaux*, Haüy stated merely that the elementary molecules of the diverse chemical principles could combine to form the constituent molecules and that the latter might have different shapes or the same shape, as in the instance of the cube. In order to have the complete solution to the problem of the structure of crystalline matter, it would be necessary to determine the figure of the elementary molecules.[1] Between 1784 and 1801 Haüy's thoughts concerning the relationship between the chemical composition and the form of the integrant molecule became clarified. Whereas earlier he had thought that one could never employ crystallography as a basis of a methodological distribution of minerals, in 1786 he stated that it appeared his theory might "furnish indications for classifying minerals in their various genera," and that from the determination of the exact figure of the integrant molecules "there ought to result knowledge with respect to the basis and the nature of substances."[2] Not only did the elementary molecules unite in such a way that they imparted distinctive chemical and physical properties to a substance, but further, their union resulted in a geometric arrangement that Haüy called the structure. The structure, he said in 1792, "is a fixed and invariable point relative to all bodies of the same species. The number and disposition of the faces which determine the form, the coloring principles and the different mixtures which modify the results of chemical analysis, all oscillate around this fixed point."[3] Moreover, the species is given immediately by nature.

---

[1] Haüy, *Essai*, p. 36.
[2] "Lettre de M. l'Abbé Haüy à M. de Lametherie sur le schorl blanc," *JDP*, XXVIII (1786), 64.
[3] R. J. Haüy, "De la structure considerée comme caractère distinctif des minéraux," *JHN*, II (1792), 65.

In botany, for example, it is the reproduction of the individuals which properly determines the species, and it is in consequence of this reproduction that all the individuals of the same species are similar in all their parts, with the exception of small and accidental differences such as height, color, and so on, which give variety.[4]

Inasmuch as each part of a mineral is always the same mineral, it was permissible to classify it as a species rather than to treat each specimen as an individual.

Thus, Haüy transferred the notion of the fixity of biological species, which was paramount in his time, to the mineral kingdom. In the 1801 edition of his *Traité de minéralogie*, he clearly stated his concept that the elementary molecules of a substance combined in certain definite proportions, and he identified the resulting chemical molecule with the integrant molecule that resulted from cleavage.[5] He postulated that without doubt the elementary molecules had regular and constant forms for each species of acid and alkali and that these adapted themselves in one way or another to form the integrant molecule. The shape of the integrant molecule depended upon the order of arrangement of the elementary molecules and their relative quantities. If the relative quantities varied, it was only natural that the new assortment would cause a change in the figure of the integrant molecule. Thus, in Haüy's view, a mineral species should be defined not only in terms of the relative proportions of its chemical constituents but also with respect to the form of its integrant molecule. He admitted that in some substances where the form of the integrant molecule was a regular geometric solid, such as a cube or a regular octahedron, the form was of little value in the definition of a species. In these instances, chemical and physical tests must be made in order to identify it properly. But, since the majority of integrant molecules could be distinguished by the values of their facial and interfacial angles, the form was a specific in these substances. "This form in turn influences the properties of the body of which it is a physical element, such as refraction and specific weight, which pertain to the essence of the body."[6] In the 1822 edition of his *Traité de cristallographie*, he attempted to demonstrate how a cubic integrant

---

[4] R. J. Haüy, "Exposition de la théorie sur la structure des cristaux," *ADC*, XVII (1793), 225.

[5] Haüy, *TDM1*, I, 5-6.

[6] Haüy, *TDM2*, I, 23.

molecule might be formed in several ways by the juxtaposition of variously shaped elementary molecules. The sections through two such cubes are shown in figure 14.[7]

Fig. 14. Haüy's conception of how variously shaped elementary molecules might combine to form identical cubic integrant molecules. Source: *R. J. Haüy,* Traité de cristallographie *(3 vols.; Paris, 1822), pl. 69, figs. 13, 14.*

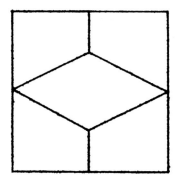 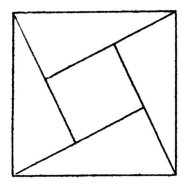

Haüy's confidence in the importance of form as a necessary ingredient in the definition of a mineral species was by no means based entirely upon his theoretical considerations. Prior to chemical analysis, he had asserted, on the basis of his analysis of the forms of the integrant molecules, that so-called Spanish chrysolite was in actuality apatite and that beryl and emerald belonged to the same species.[8] Thus, the determination of structure was a type of analysis solely within the province of crystallographic mineralogy. Although the analysis did not proceed so thoroughly as the analytical methods of chemistry, it could be employed in many circumstances as a characteristic to aid in distinguishing a particular mineral. Structure analysis offered the advantage of simplicity in that it could be accomplished in most instances more easily than chemical analysis, and, when chemically complicated minerals were under consideration, it sometimes yielded a more distinct result. In Haüy's opinion, chemistry clarified the qualities and relative quantities of the elementary molecules, but imparted no knowledge

---

[7] Haüy, *TDC*, II, 429.
[8] Haüy, *TDM1*, II, 244, 528.

of their functions. Crystallography repaired this defect; it made known the form of the integrant molecule which had been imprinted with the functions of the elementary molecules. The alliance of the two sciences was indispensable in order to have an exact and complete notion of the species. To the objection that there was no proof that the true integrant molecules of substances were known, Haüy replied that it was not certain that chemistry had penetrated to the true elementary principles of bodies. He believed the mechanical division of crystals and their chemical analysis yielded so exact a definition of species that future advances in the understanding of nature would not invalidate the deduction of a species resulting from these contemporary research methods. Minerals had a geometric type consisting of the form of the integrant molecule and a chemical type grounded in the composition of the integrant molecule; the definition of species was founded upon the coexistence of these two types in an individual specimen. Such an approach, Haüy believed, was the only logical method by which minerals could be classified systematically.[9]

Haüy prepared a mineral classification system of classes, orders, genera, and species, with the four large class divisions being acidiferous, earthy, nonmetallic combustible, and metallic substances. Although giving lip service to the importance of chemical composition, Haüy actually relied more upon the form of the integrant molecule in delineating the place of a mineral in his classification. For example, noting that the integrant molecule of calcium-manganese-iron carbonate was the same as that of calcium carbonate (calcite), he placed this "chaux carbonatée ferrifère avec manganese" as a subspecies of calcium carbonate, even though Bergman had reported that it contained 38 percent iron oxide and 24 percent manganese oxide.[10] Adhering strictly to his concept of species and not suspecting the phenomenon of isomorphism, Haüy explained that the calcium carbonate had accidentally entrained iron during its crystallization in such a manner that the molecules conserved their tendency to arrange themselves according to laws peculiar to them.

But despite Haüy's meticulous work in discriminating between the integrant molecules and the lengthy annotations that he added to justify his classification, many chemists refused to credit the

[9] Haüy, *TDM2*, I, 25-26.
[10] Haüy, *TDM1*, II, 177-184.

form of the integrant molecule with the importance Haüy attached to it. The attitude was expressed in this statement:

It is by comparison of many chemical analyses of the same mineral that the species to which it belongs can be ascertained. Consequently, all classification that one can make with the aid of physical characteristics and particularly form, in waiting to unite enough facts relative to the composition for a definitive determination, may be regarded as merely provisional. . . . the principle that minerals in which the form of the molecules, or rather the cleavage, is different are different species is not a law of nature; it is a rule of convenience which may have exceptions. . . . it is the composition that establishes the essence of a mineral; the form is, after all, only a means of recognition.[11]

Referring to those substances whose integrant molecules were cubes, regular tetrahedra, or regular octahedra, Berthollet objected that it was unreasonable to give such confidence to a characteristic that did not indicate any difference between two substances of differing compositions, where at the same time very slight differences in chemical composition produced a change in form.[12]

Haüy's species concept, Berthollet thought, led to useless divisions and did not include numerous crystalline minerals whose chemical compositions were extremely complicated or whose integrant molecules were incapable of determination. The noted British chemist, Thomas Thomson (1773-1852), echoed Berthollet's views, stating that a rigid definition of mineral species, as Haüy had attempted, was impossible.[13] He postulated that an identity of chemical composition of two minerals might exist despite a difference in the shapes of their integrant molecules, so that they might constitute the same chemical species but different species from a crystallographic point of view. Haüy's species were unconnected generically, he said, and thus his system of classification constituted nothing more than a catalog. Further, Thompson was particularly incensed that Haüy had, in the process of proposing his mineralogical system, attempted to reform the nomenclature as well, and he objected strenuously that Haüy had discarded all the common and well-known mineralogical terms and had introduced, instead,

[11] "Lettre de M. d'Aubisson à M. Berthollet," *ADC*, LXIX (1809), 176, 180.
[12] C. L. Berthollet, *Essai de statique chimique* (2 vols.; Paris, 1803), I, 433-449.
[13] Thomas Thomson, "Some Observations in Answer to Mr. Chenevix's Attack upon Werner's Mineralogical Method," *AP*, I (1813), 241-258.

some one hundred words of his own invention. It should be noted, however, in fairness to Haüy, that his structural classification, rather than the chemical classification of his critics, conforms closely to modern practice.

In addition to such objections to Haüy's concept of mineral species and to his system of classification, other chemists believed that his geometric construction of a particular crystal variation by the assignment of certain laws of decrement to a primitive form presented only an indistinct outline of a complicated phenomenon. One could grant that some order had been brought out of the confusion existing previously, but there were many important unanswered questions. Why were the integrant molecules of one substance tetrahedra and those of another rhombohedra? Although it was known that the conditions of crystallization influenced the final crystal form and that the presence of foreign ingredients in a solution had a similar effect, no definite reasons could be given as to why nature in one case produced one possible crystal variation of a substance rather than another. What were the influences at work which gave rise to the manifest complexity? Haüy had removed the consideration of such questions in his treatment of crystalline matter:

My design is not to search for the manner of action of the primitive forces of crystallization. I do not know whether it would be possible to find all the factors that enter into such a theory, such as the volume of the molecules on which the existing forces exercise their action, the density of the fluid, its temperature, and so forth, which influence the formation of the crystal and which it would be necessary to submit to calculation to resolve problems of this nature completely.[14]

Some chemists believed, however, that careful chemical experimentation ought to shed some light on the mechanism of the process of crystallization. This represented an entirely different intellectual approach.

One of the early workers in this area of research was Nicolas Leblanc. Leblanc is remembered chiefly as the discoverer, in 1790, of the important process for the economical manufacture of soda ash, the utilization of which gave a powerful stimulus to progress in the soap and textile industries of Europe. His life is a record of constant misfortune. It is doubtful that he collected the prize

---

[14] Haüy, *Essai*, p. 8.

money offered by the Academy of Sciences for the invention of a successful soda ash process. During the revolution, the government exerted pressure on him to make his patents public property, a move that caused his financial failure. Under the Empire, his patent rights were restored, but capital promised to aid in reopening his factory was not forthcoming. He occupied a few minor civil posts and was respected by contemporary scientists, but his continued destitute condition caused despondency, and he committed suicide in 1806.[15]

Leblanc was intensely interested in the mechanism of crystallization; in fact, his method of recovering soda ash involved successive crystallizations of solutions of sodium carbonate. He believed that the true science of crystallography should encompass the study of the conditions of crystal growth so that the production of a particular crystal form might not only be explained but predicted.[16] Science and art should meet in "cristallotechnie," Leblanc's term for this research. Scientists ought to be able to produce a crystal variation at will and modify it. This technique was an art in that the processes of nature would be duplicated, and a science because the laws of causality would be verified in the process.

Leblanc believed that crystals were susceptible to two kinds of variation. First, substances crystallized in two entirely different forms, for example, alum as a cube or an octahedron. Second, there were accidental and superficial variations of the same form; for instance, the corresponding faces of two octahedral alum crystals could be proportionately larger or smaller. In order to account for these two types of variation, Leblanc thought it necessary to study three kinds of phenomena. The chemical composition itself was of the utmost importance. His research indicated that an excess of acid or of base might cause a change in the crystal form such that in many salts there could be two types of combination which could result in two different crystal forms. Further, Leblanc noticed that many sulfates combined among themselves in all proportions. Copper sulfate, for example, either alone or in a mixture of iron

---

[15] *Biographie universelle (Michaud)* (52 vols.; Paris, 1853-1866), XXIII, 464-466. See also C. C. Gillispie, "The Discovery of the Leblanc Process," *Isis*, XLVIII (1957), 152-170.

[16] N. Leblanc, "Essai sur quelques phénomènes relatifs à la cristallisation des sels neutres," *JDP*, XXVIII (1786), 341-345; and "De la Cristallotechnie ou Essai sur les phénomènes de la cristallisation," *JDP*, LV (1802), 296-313.

sulfate, produced crystals that had a rhombohedral configuration, although there was some difference in their color. In addition, one should study the physical condition of the solution because this influenced the crystal figure. Leblanc believed the force of attraction was the efficient cause of the union of the saline molecules, but that this force was subject to the qualifying action of the force of gravity. Hence, the position of the crystal nucleus in the solution influenced the resulting crystal form. The sizes of the faces, he thought, depended upon the relationship existing between the length and the width of the growing nucleus. This ratio changed depending upon whether the nucleus was at the top, the middle, or the bottom of the crystallizing solution. Finally, there were external factors that came into play to cause variations in the crystal form. Humidity, temperature, and the movement of air currents provoked changes in the solution and consequently in the eventual crystal form. According to Leblanc, all these factors had one end result: they caused a new set of conditions to prevail in the evaporating liquid. The crystals that were the end products bore the marks of the changing conditions; they were not mere end effects of simple gradations from the primitive forms as they had been treated by Haüy.

Leblanc's analysis of the factors that influenced the crystallization process was perceptive. In his study of copper and iron sulfate mixtures, one can see a vague notion of the existence of solid solutions, but Leblanc did not pursue his research far enough to arrive at this result. More than a decade passed before any further important studies of the mechanism of the crystallization process were conducted. In 1815 the German chemist, Johann Fuchs, in reporting on the properties of a newly found mineral, gehlenite, a silicate of calcium and aluminum, noted the presence of iron oxide. In commenting on this, he stated:

... the iron oxide is not an essential constituent of this combination but merely a vicarious constituent, if I may use this expression, for the substitution of almost just as much lime, which in the absence of the iron oxide must be present as a supplement in order to combine with the other constituents in the proper proportions; and I believe that in the future varieties will be found which will have much less or no iron oxide.[17]

---

[17] J. Fuchs, "Ueber den Gehlenit, ein neues Mineral aus Tirol," *SJ*, XV (1815), 382-383.

As further proof of such substitutional activity, he pointed out that ammonia might replace potassium to some extent in its sulfate, and in 1817 he stated that the similarity of the crystal forms of the sulfates of calcium, strontium, and lead might be explained in this way.[18]

Simultaneously, François Sulpice Beudant, a former student of Haüy's, embarked upon an ambitious program, actually one outlined by Leblanc, and he published his findings in 1818.[19] Beudant gave full credit to Leblanc's work and stated he was using it as a basis for his own research. Although he paid his respects to Haüy, he remarked that however satisfying Haüy's system was, there remained a thick veil over the causes of crystal variations; there were lacunae in the science of crystallography. Beudant was convinced that the occurrence of diverse crystal forms of the same substance was not mere chance. He pointed out that the same crystal forms of a mineral were found in different localities where analogous geological conditions prevailed. Conversely, where the earth strata and neighboring minerals were different, a different crystal form of the same substance was seen. He believed that this complicated problem could be solved in the chemistry laboratory where all variable conditions might be simulated.

Beudant's numerous experiments were designed to investigate four sets of variables: (1) the effects of circumstances external to the solution itself, (2) the effects of the introduction of foreign matter into the solution which did not combine chemically with the constituents, (3) the effects of the variation of the relative proportions of the constituents, and (4) the effects of the presence of other chemicals in the solution which might or might not combine with the principal constituents. Temperature, of course, was the principal external influence since it caused more or less rapid evaporation. When evaporation was rapid, the crystals were small, badly formed, and often confused. Lack of humidity caused the growth of crusts. But change of temperature or humidity and consequent rapid or slow evaporation did not produce a change in the crystal form of the same substance. Similarly, variation of the barometric pressure or the degree of concentration of the solution did not cause any marked change of form, although Beu-

---

[18] J. Fuchs, "Ueber einige phosphorsauere Verbindungen," *SJ*, XVIII (1816), 288-296.

[19] F. S. Beudant, "Recherches sur les causes qui peuvent faire varier les formes cristallines d'une même substance minérale," *ADC*, VIII (1818), 5-50.

dant agreed with the findings of Joseph L. Gay-Lussac (1788-1850) that crystallization in a vacuum did produce radical changes in the crystal form of some substances.[20] Changes in the volume, height, or state of electric charge caused variations in the size of the crystals but not in the form. A greater volume of solution yielded larger crystals, whereas smaller crystals were formed when the solution was charged with electric fluid. He agreed with Leblanc's conclusion that crystals growing close to the bottom of the solution tended to become wider than those growing at the middle or top of the containing vessel. The external variable circumstances, then, although influencing the perfection and the size of the crystal, had no substantial effects on the basic crystal habit.

Similarly, he found that suspended particles, precipitates, or gelatinous substances introduced into the crystallizing solution exercised little influence on the form of the resulting crystals. In some instances the foreign particles were entrained in the growing crystal and caused a slight modification in the appearance of the faces. Gelatinous substances seemed to retard the evaporation of the fluid so that more perfect crystal forms were produced. Sediments appeared to have no effect at all. With respect to the relative proportions of the constituents, however, Beudant found that an excess of acid or base in the solution did produce noticeable variations of the form. He confirmed Leblanc's observation that a larger proportion of the basic constituent in crystallizing solutions of alum resulted in the formation of cubic rather than octahedral crystals. Iron sulfate crystallized in more complicated forms when an excess of acid was present, whereas the crystals of copper sulfate were more simple under the same circumstances. He concluded that the proportion between the acid and the base exercised a considerable influence on the eventual crystal form and that marked effects could be noticed with just a minor variation in the relative proportions.

While engaged in this research, Beudant turned his attention to crystallizing solutions containing varying amounts of the sulfates of iron, copper, and zinc, the problem that had interested Leblanc. Beudant wished to determine up to what point a combination in definite proportions, such as copper sulfate, might admit other chemical principles without changing its crystalline form. The

[20] J. L. Gay-Lussac, "De l'influence de la pression de l'air sur la cristallisation des sels," *ADC*, LXXXVII (1813), 225-236.

results of this research were most perplexing.[21] Haüy had stated that the integrant molecule of iron sulfate was an acute rhombohedron, that of copper sulfate an irregular oblique-angled parallelepiped, and that of zinc sulfate probably a regular octahedron. Beudant found that solutions containing 85 percent zinc sulfate and 15 percent iron sulfate, or those containing 90 percent copper sulfate and only 9 to 10 percent iron sulfate, still yielded crystals having the same rhombohedral form of pure iron sulfate. Even solutions of 97 percent mixed zinc and copper sulfate and 3 percent iron sulfate produced the rhombohedral crystals.

There were several conclusions to be drawn from these results. Beudant stated that a definite chemical combination, that is, the iron sulfate, could, without the crystal system proper to it being changed, admit foreign principles not only up to a noticeable proportion of its weight but up to thirty-two times greater. There could exist a constituent in a chemical substance which might be present only in a very small quantity, but which, far from being regarded as accidental, exercised an extremely strong influence on the properties of the substance since it could produce the form characteristic of its crystal system. This, of course, posed a problem with respect to the classification, for example, of the salt containing 91 percent copper sulfate and 9 percent iron sulfate. If one classified it with the sulfates of copper, one would have a salt possessing the form of the sulfate of iron. If it was classified according to crystal form, one would have among the sulfates of iron a salt containing 91 percent copper.

But there were other implications of the findings, Beudant thought. They proved that the sulfates of zinc and copper were not present in these salts in definite proportions and that consequently, according to the principles embraced by the majority of contemporary chemists, these composite salts were not combinations but mixtures. Such chemical mixtures (*mélanges chimiques*) must be presumed to exist in a large number of minerals in which the quantities of the elements present did not correspond to what should be expected in a combination in definite proportions. Beudant did not wish to espouse Berthollet's doctrine that many compounds could have variable compositions, but at this

[21] F. S. Beudant, "Recherches tendantes à déterminer l'importance relative des formes cristallines et de la composition chimique dans la détermination des espèces minérales," *ADC*, IV (1817), 72-84.

point he was puzzled as to the exact nature of his chemical mixtures.

The Academy of Sciences directed Haüy, Brochant de Villiers, and the noted chemist, Louis Nicolas Vauquelin (1763-1829), to investigate and report on Beudant's findings. This committee accepted Beudant's conclusions but added that the experiments with the mixtures of the sulfates demonstrated that, in some chemical mixtures, one component directed or had a dominant influence in the production of a crystal form from the solution to the exclusion of the other constituents. In summarizing the report, Cuvier, the permanent secretary, stated that iron sulfate apparently exercised an astonishing despotism in this respect. He believed a great deal of investigation would be necessary to explain how so small a number of rhombohedral integrant molecules could arrange themselves on similar faces to form the rhombohedral mixed crystals without being troubled by the prodigious number of integrant molecules of other figures.[22]

William Hyde Wollaston had a different suggestion. Previously, in 1812, he had written a memoir objecting that Haüy had assigned the same primitive form to three spars, that is, carbonate of lime, magnesium carbonate, and iron carbonate, whereas his measurements demonstrated that the corresponding angles differed slightly in their values. At that time he said it was

... very evident from the numerous analyses that have been made of iron spar by other chemists how extremely variable they are in their composition, and consequently how probable it is, that the greater part of them are to be regarded as mixtures; although it be also possible, that there may exist a triple carbonate of lime and iron as a strict chemical compound.[23]

Here Wollaston clearly suggested solid solution, but, answer to Beudant's memoir, he was more explicit. He insisted that the transparency of the mixed crystals ruled out the possibility of a mere mixture, that "some more intimate chemical union may be presumed to occur." He had, he said, produced crystals from a

---

[22] G. Cuvier, "Analyse des travaux de l'Académie royale des sciences pendant l'année 1817, partie physique," *ADS* (1817), p. ciii; and "Analyse des travaux de l'Académie royale des sciences pendant l'année 1818, partie physique," *ADS* (1818), p. cxcvi.

[23] W. H. Wollaston, "On the Primitive Crystals of Carbonate of Lime, Bitter Spar, and Iron Spar," *PT* (1812), pp. 159-162.

mixed solution of copper and zinc sulfate in the proportion of about four to one, "so nearly agreeing in the measures with those of sulphate of iron, that I cannot at present undertake to say wherein any difference consists." He concluded that crystal form under these circumstances was not a reliable indication of species. "The existence, however, of mixed crystals such as these are conceived to be by M. Beudant, cannot be questioned, and must continue to mislead those who think it possible to rely on crystalline forms alone."[24] In reply Beudant clarified his own views and protested against Wollaston's interpretation of his conclusions.

M. Wollaston appears to believe that I consider the crystals I obtained as mechanical mixtures of diverse other salts; ... I have never had that idea. ... I think also that one can, if one wishes, look upon these associations as combinations, but, since they are made in variable proportions, it is necessary to distinguish them from combinations in definite proportions, and it is for this reason that I adopted the expression *mélange chimique*.[25]

Here Beudant made a clear statement of the phenomenon of solid solution, and, together with Wollaston, he should be given credit for its recognition.

The importance of these investigations is that they led directly to the statement of the law of isomorphism. This law, proposed by Eilhard Mitscherlich in a paper to the Berlin Academy in 1819, was based primarily on his faith in the accuracy of Dalton's hypothesis and his law of chemical combination of atoms by definite proportions, although Mitscherlich professed ignorance of the shape or constitution of these atoms.[26] Dalton merely touched on the phenomenon of crystallization when he delineated his theory in 1808. He believed that crystallization exhibited the effects of the natural arrangement of the ultimate particles of the various compound substances, but he confessed that the insufficient knowledge of chemical synthesis precluded any real understanding of the rationale of this process. He supposed that the rhombohedral

24 W. H. Wollaston, "Observations on M. Beudant's Memoir 'Sur la détermination des espèces minérales,'" *AP*, XI (1818), 283-286.

25 F. S. Beudant, "Lettre au sujet du mémoire de M. Wollaston," *ADC*, VII (1817), 403.

26 E. Mitscherlich, "Ueber die Kristallisation der Salze in denen des Metal der Basis mit zwei proportionen Sauerstoff verbunden ist," *ADB* (1818-1819), pp. 427-437.

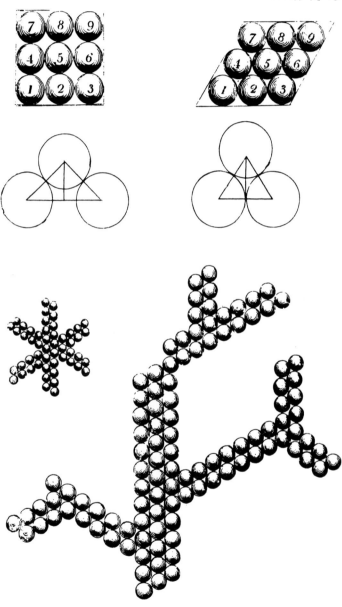

Fig. 15. *Dalton's drawings picturing the arrangement of particles of water (upper left), their arrangement in ice (upper right), perpendicular sections of a particle resting upon two others (center), the structure of ice upon sudden freezing (bottom left), and ice at the commencement of freezing (bottom right).* SOURCE: *John Dalton,* A New System of Chemical Philosophy *(2 vols.; Manchester, 1808), I, pl. 3.*

form might arise from the proper position of four, six, eight, or nine spherical particles, the cubic form from eight, the tetrahedral form from three, and the hexagonal prism from seven particles. He used sketches (fig. 15) to illustrate these possible arrangements, and, in viewing these, one is reminded immediately of the similar drawings Robert Hooke and Christian Huygens had made more than a century earlier. Dalton concluded his observations with the statement:

Perhaps, in due time, we may be enabled to ascertain the number and order of elementary particles, constituting any given compound element, and from that determine the figure which it will prefer on crystallization ... but it seems premature to form any theory on this subject till we have discovered from other principles the number and order of the primary elements.[27]

Mitscherlich had studied Oriental philology and history at Heidelberg and Göttingen and had written his doctoral dissertation in the field of Persian history. He became disillusioned with these disciplines, however, when he was unable to find a suitable position, and he turned to the study of medicine. He became intensely interested in chemistry while pursuing his medical education and was attracted particularly by the ideas of Dalton and Jons Jakob Berzelius (1779-1848). This brilliant Swedish chemist had conceived the notion of using oxygen as a basis of reference for the determination of the atomic weights of the other elements with which it combined, and by 1818, when Mitscherlich began his research in Berlin, Berzelius had determined and published the combining proportions of oxygen and other elements in many substances. Mitscherlich began his work with the study of phosphorus and arsenious salts in which oxygen was combined in the same proportions and which had identical amounts of water of crystallization. He determined that the combination of these salts with the same metallic bases resulted in the production of crystals of the same figure. He extended his research and found that the oxides of magnesium, copper, and nickel, when combined with the same acid, crystallized in the same configuration. At this point he decided to repeat Beudant's experiments on the sulfates of manganese, copper, iron, zinc, magnesium, cobalt, and nickel. He dis-

---

27 John Dalton, *A New System of Chemical Philosophy* (2 vols.; Manchester, 1808), I, 211.

covered that the same crystal form was produced in these sulfates when they had the same proportionate amount of water of crystallization. On the basis of these results, Mitscherlich felt confident in announcing in his first paper to the Berlin Academy that a chemical combination in which the number of atoms was the same took the same crystal form. When two acids were combined in the same proportions with the same base, the products of crystallization would have the same form, and, similarly, the crystal figures resulting from the union of two bases with the same acid in the same proportion would be identical. The crystal form, then, was independent of the chemical nature of the atoms. Thus, Mitscherlich's paper contained the first definitive statement of isomorphism, and he was aware that some modification might have to be made in the future.

Berzelius had been invited to Berlin in September, 1819, to discuss an offer of the chair of chemistry at the University of Berlin which had been vacated by the death of Martin Klaproth (1743-1817). He met Mitscherlich and discussed the content of the memoir Mitscherlich had prepared for presentation to the academy. Berzelius was so impressed with Mitscherlich's work that he recommended his appointment to the vacant chair. As Mitscherlich was then only twenty-five years of age, he was considered too young for the post, and he was sent to work with Berzelius at public expense. Upon his return from Sweden in 1822, he was appointed extraordinary professor of chemistry at Berlin and three years later assumed the chair as full professor.[28]

Berzelius considered the work of Mitscherlich and the contemporaneous discovery in 1819 by the two French physicists, Pierre Louis Dulong (1785-1838) and Alexis Thérèse Petit (1791-1820), that the specific heats of solids varied inversely as their atomic weights, to be the two most important empirical proofs of the atomic theory. Almost immediately, the presence of isomorphism came to be used as a basis for the determination of atomic weights. If two substances had similar crystal forms and there was reason to believe they had analogous chemical compositions, a relatively simple mathematical procedure could be used to calculate the atomic weight of an element that was a constituent of one isomorphous chemical compound when the atomic weight of

[28] A. Mitscherlich, ed., *Gesammelte Schriften von Eilhard Mitscherlich* (Berlin, 1896), pp. 1-120.

the analogous element in the other compound was known. Mitscherlich, for example, determined the atomic weight of selenium by a study of potassium sulfate and potassium selenite, which he declared to be isomorphous. The quantities of sulphur and selenium which were equivalent in these substances were considered to represent the total weight of an equal number of atoms and should therefore be in the ratio of their atomic weights. Since the atomic weight of sulphur was known from other determinations, the atomic weight of selenium could be calculated readily.

Yet, in order to gain the full advantage of this procedure, it was necessary that the chemical compounds investigated be pure and in definite proportions; there could be no consideration of mixed crystals of the type Beudant and Wollaston had discussed. As a result, as C. S. Smith has noted, it was the molecule as the combination of atoms in fixed proportions—rather than the atoms themselves—which became the center of chemical interest.[29] The attention of chemists was diverted from the atom as such, and, along with this, from solid solutions, or what is now termed ionic substitutions. The ideas of Berthollet, who recognized that many compounds could have variable compositions and who insisted that chemical reaction depended upon the relative amounts of the constituents, as well as their affinities, were pushed into the background. The implications of Beudant's and Wollaston's recognition of solid solutions were left undeveloped for the time being.

If applied strictly, Mitscherlich's law of isomorphism was in direct opposition to Haüy's insistence that the form of the integrant molecule was a specific characteristic of a substance. Berzelius attempted on a number of occasions to win Haüy's acceptance of Mitscherlich's findings, but without success.[30] Haüy adamantly maintained that the form of the integrant molecule was a specific, related to the same chemical combination whenever the molecule was not a "limit" form, that is, a cube, a regular tetrahedron, or a regular octahedron. He wrote that he had discussed the matter with the most distinguished scientists in France who agreed that Mitscherlich was in error. He argued that Mitscherlich had been badly deceived in identifying forms whose corresponding

[29] Cyril Stanley Smith, *A History of Metallography* (Chicago, 1960), pp. 190-193.

[30] "Correspondance d'Haüy et de Berzélius," *Bulletin de la Société Française de Minéralogie*, LXVII (1944), 199-202.

angles had values as much as 10 degrees apart. In this regard, it is true that Mitscherlich very shortly had to replace the notion of the identity of the forms of isomorphic substances with that of their striking similarity. Further, Haüy declared that the results of Mitscherlich's experiments were questionable because they proceeded from laboratory experiments on artificial salts. These procedures, Haüy stated, were not analogous to the processes by which the minerals of the terrestrial globe had been formed. Mitscherlich, according to Haüy, had confused laws with conditions. Laws were the same always; they operated in nature, and everything in nature depended upon them. Results subject to prior planned sets of conditions or experiments, however, could vary endlessly.[31] But to Brochant de Villiers, Haüy confessed that "if Mitscherlich's theory is correct, mineralogy would be the most wretched of the sciences."[32] Berzelius finally recognized that Haüy could not be convinced of the validity of the law of isomorphism:

One ought not expect that a gray-haired scientist close to the end of an honorable life should give up, without resistance or any attempt at justification, a theory whose declaration he erroneously considered to be the most important of his discoveries; this is perhaps too much to morally demand of any man.[33]

But still there were some who continued to insist upon the validity of Haüy's view. Beudant, for example, maintained rightly that isomorphous substances were merely similar rather than identical in form. And with respect to the forms being indicative of species, he said that isomorphous substances lend inexactness but they "cannot destroy the notion of species. . . . just because there exist mulattos and quadroons, and so on, no one imagines that this is a reason for not being able to distinguish the Negro races and the white races."[34] Despite such arguments, however, the discovery of isomorphism was a severe blow to Haüy's theory.

Mitscherlich's later studies of the form of crystalline substances provided the solution to another problem that had concerned crys-

---

[31] A. Mitscherlich, op. cit., p. 173n.
[32] Edward H. Kraus, "Haüy's Contribution to Our Knowledge of Isomorphism," American Mineralogist, III (1918), 129.
[33] A. Mitscherlich, loc. cit.
[34] F. S. Beudant, "Sur la classification des substances minérales," ADC, XXXI (1826), 185.

tallographers and chemists for more than twenty years. The discussion focused on calcite and aragonite which had completely dissimilar integrant molecules but apparently the same chemical composition. In 1822 an American mineralogist wrote: "The analysis of no mineral has ever so much exercised the talents, exhausted the resources, and disappointed the expectations of the most distinguished chemists in Europe, as that of arragonite."[35] Romé de l'Isle had designated aragonite as a variety of prismatic calcite which was found in Spain, and Klaproth, with the purpose of ascertaining whether Romé de l'Isle's statement was correct, had analyzed the mineral and found that it was indeed composed of lime and carbonic acid.[36] Werner had named the substance "aragonite" and had listed it as a completely separate species because of the marked differences in specific weight, hardness, and external form between it and calcite. When Haüy was preparing the first edition of his mineralogical treatise, he measured the primitive form of aragonite and found that it was an irregular hexahedral prism with interfacial angles of 116° and 128°, and he determined that this was an assemblage of four prismatic integrant molecules.[37] On the basis of this finding, as well as the fact that its physical properties differed from those of calcite, he judged that this form could not be a variety of the integrant molecule of calcite, and he listed it as a separate species, but, nevertheless, as one on which further observations should be made.

Haüy's action of eliminating aragonite from the species of calcium carbonate stimulated a rash of experiments designed to produce a perfect analysis as a basis for its inclusion in this mineral species. In 1804 Vauquelin and Fourcroy determined that the relative proportion of lime and carbonic acid in calcite was 57 to 43, whereas their proportion in aragonite was 58.5 to 41.5. They found no other chemical ingredients in either substance and did not believe that the slight difference in the proportions of lime and carbonic acid was significant.[38] In 1807 Louis Jacques Thénard

35 Parker Cleaveland, *An Elementary Treatise on Mineralogy and Geology* (Boston, 1822), p. 193.

36 J. B. L. Romé de l'Isle, *Cristallographie, ou description des formes propres à tous les corps du règne minéral* (4 vols.; Paris, 1783), I, 517.

37 Haüy, *TDM1*, IV, 338-339.

38 L. Vauquelin and A. Fourcroy, "Expériences comparées sur l'arragonite d'Auvergne et le carbonate de chaux d'Islande," *AMHN*, IV (1804), 405-411.

(1777-1857) and Jean Baptiste Biot (1774-1862) presented similar analytical results:

	Calcite	Aragonite
Lime	56.351	56.327
Carbonic acid	42.919	43.045
Water	.730	.628
	100.000	100.000

Because the difference in analysis was not considerable, they suggested that either the elementary molecules, the chemical principles, had the faculty of combining together in many ways or they acquired this capability by the influence of a transient element that disappeared without destroying the combination.[39]

Despite the results of these and many other analyses by noted chemists, Haüy was reluctant to admit that the form of the integrant molecule did not demonstrate aragonite as a completely separate mineral species. Rather, he and his followers, adhering to their definition of species, emphasized the importance of the crystal form and such other physical tests as hardness and specific weight as being the important identifying characteristics of species.[40] De Bournon stated that aragonite offered a striking example of the necessity at times of abandoning chemical analysis in the determination of a species in favor of these other characteristics. It was impossible to conceive of two substances with the same chemical properties but whose specific weights were so different. He suggested there must be another chemical principle present in aragonite which occasioned the great difference in form and physical properties. According to de Bournon, a mineral substance could not be present because it would undoubtedly have been found by these careful analyses. If, however, the foreign principle was one of the fluid elements of nature, such as caloric, the fluid of light, or that of electricity, it would not be astonishing that the contemporary techniques of chemistry had been unable to find it.[41]

De Bournon agreed with Haüy that there was no possibility of

[39] J. B. Biot and L. Thénard, "Mémoire sur l'analyse comparée de l'arragonite et du carbonate de chaux rhomböidal," *GJ*, V (1808), 237-242.
[40] R. J. Haüy, "Sur l'arragonite," *AMHN*, XI (1808), 241-270.
[41] J. L. Bournon, *Traité complet de la chaux carbonatée et de l'arragonite* (3 vols.; London, 1808), II, 119-147.

any law of decrement which would link calcite and aragonite. But Johann Jacob Bernhardi believed that he could prove such a connection existed. He supposed certain complicated laws of decrement which operated on only four of the edges and two of the angles of the calcite integrant molecule and which would produce the usual crystal form of aragonite. It seemed to Bernhardi that the difference in the optical properties of calcite and aragonite stemmed from a difference in the polarity of the light axes, and that possibly the cause of the diverse specific weights and hardness should be sought in further studies of this property.[42] Haüy was disturbed by Bernhardi's arbitrary choice of certain nonsymmetrical edges and angles as bases on which the decrement operated, and he demonstrated that the cleavage of aragonite did not support the proposed laws. Further, he checked Bernhardi's results and determined that one interfacial angle was produced that was 1° 8′ different from the measurement of the corresponding angle of the aragonite crystal. He did not believe that in this instance a discrepancy could be admitted. In addition, Haüy had by this time determined that the integrant molecule of aragonite was an irregular octahedron, one so different from the calcite rhombohedron that no geometrical relationship could exist between the two. He was firm in his assertion that aragonite was a completely different species.[43]

In 1813 an event occurred which apparently gave a signal victory in the controversy to Haüy. Frederick Stromeyer (1766-1835), professor of chemistry at Göttingen, under whom Mitscherlich studied, announced that aragonite contained strontium carbonate which he had not found present in calcite. This carbonate, he thought, was chemically united with the carbonate of lime in a constant proportion and constituted a natural triple combination of carbonic acid with lime and strontium. The quantity of strontium carbonate in the aragonite sample he analyzed, according to Stromeyer, was between 3 and 4 percent. He attributed the fact that it had not been recognized previously to the assumption that any strontium sulfate would be precipitated from a solution of hydrochloric or nitric acid by the addition of sulfuric acid. Stromeyer stated that there was so strong a resemblance in the chem-

---

[42] J. Bernhardi, "Beweis dass die Form des Arragonits aus der Grundform des Kalkspaths abgeleitet werden könne," *GJ*, VIII (1809), 152-162.

[43] R. J. Haüy, "Addition au mémoire sur l'arragonite," *AMHN*, XIII (1809), 241-253.

ical properties of strontium and lime that it was difficult to separate them. He accomplished this task by dissolving the aragonite
in pure nitric acid, evaporating the solution to crystallization, and
then treating the crystalline mass with alcohol in which strontium
nitrate was not soluble. To the question of whether the quantity
of strontium carbonate he had found was sufficient to produce the
striking peculiarities of aragonite, Stromeyer answered in the affir-
mative. He supposed that the crystal form of aragonite depended
upon a dominant power possessed by strontium carbonate which
caused the calcium carbonate to crystallize in an entirely different
form from calcite.[44] This faculty was the same as that which the
French academicians several years later attributed to iron sulfate
in Beudant's experiments.

When Haüy related the results of Stromeyer's experiments to
Vauquelin, the latter attempted to duplicate the findings. After a
series of unsuccessful analyses, he was finally able to detect 0.6
percent of strontium in one sample of aragonite. He was not convinced that this quantity could produce so marked an effect on the
crystal form.[45] But Haüy, believing that Stromeyer's results should
receive further consideration, requested André Laugier (1770-
1832) to repeat Stromeyer's experiments. Laugier followed Stromeyer's analytical method in detail and obtained the predicted
strontium nitrate. Agreeing with Stromeyer that French chemists
had overlooked the presence of strontium carbonate in aragonite,
he attributed their oversight to their use of alcohol that was not
sufficiently pure. It contained enough water to dissolve the small
amount of strontium nitrate which was present. Laugier's cor-
roboration of Stromeyer's findings apparently solved the problem
of aragonite and vindicated Haüy's judgment that it was a separate
species.[46] One announcement of Laugier's results included this
remark:

Had not the father of crystallographic science so clearly and firmly
maintained the diversity of primitive structure in arragonite and car-

---

[44] Anonymous, "Discovery of the Composition of Arragonite," PM, XLII
(1813), 25.
[45] L. Vauquelin, "Pour déterminer les rapports de l'acide carbonique dans
les carbonates de chaux, de baryte, de strontiane, dans l'arragonite, le cuivre
bleu, et le cuivre vert de Chessy, suives de l'analyse de l'arragonite d'Auvergne," ADC, XCII (1814), 311-319.
[46] A. Laugier, "Note sur la présence de la strontiane dans l'arragonite,"
Mémoires de Museum d'Histoire Naturelle, I (1815), 66-68.

bonates of lime, chemists would not again have thought of looking for any dissimilarity in their chemical constitution, after the number of respectable analysts who have pronounced their identity.[47]

It was soon determined, however, that the correct answer to the problem had not been provided. Analyses of samples from various localities showed varying amounts of strontium carbonate in aragonite. Crystals of strontium carbonate were found in Germany and were reported at first to be strikingly similar in form to those of aragonite, but when they were sent to Haüy, he determined that the interfacial angles of the primitive forms of the two substances were sufficiently different that they could not be considered identical. Futhermore, no law of decrement could relate one form to the other. In publishing the results of this crystal analysis, Haüy admitted that the conclusion destroyed all the value of the chemical experiments that had resulted in the discovery of strontium carbonate in aragonite. If the integrant molecules of strontium carbonate and aragonite were different, one could not reasonably attribute the form of aragonite to the presence of a small amount of strontium carbonate.[48] In the 1822 edition of his *Traité de minéralogie*, therefore, Haüy admitted the possibility that a similar composition might give rise to integrant molecules of entirely diverse forms which had dissimilar physical properties. The difference could depend upon the mode of existence of the elementary molecules at the time of the formation of the integrant molecules of the two variations. It appeared that this anomaly occurred in the case of the diamond and carbon which Thénard and Sir Humphry Davy (1778-1829) had certified to be chemically identical. It could be true that aragonite offered another example of the same phenomenon. However, Haüy remarked, these examples should not be construed as vindications of the superiority of chemical analysis in the determination of a mineral species. On the contrary, these examples proved that chemistry did not offer a positive method of identification. The forms, hardness, specific weights, and optical phenomena of calcite and aragonite were completely different, and these differences attested to the fact that these two minerals should be considered different species.[49]

---

[47] Anonymous, "Strontian in Arragonite," *PM*, XLV (1815), 390.
[48] R. J. Haüy, "Comparaison des formes cristallines de la strontiane carbonatée avec celles de l'arragonite," *ADC*, V (1817), 439-441.
[49] Haüy, *TDM2*, I, xviii, 477-487.

Mitscherlich had extended his studies of the phenomenon of isomorphism while he was working with Berzelius in Sweden. In 1821 he presented a memoir to the Academy of Stockholm in which he gave the results of many new experiments to support his theory of isomorphism.[50] At the end of the paper, he considered the crystal forms of aragonite and calcite. He agreed that these substances crystallized in two entirely different forms, but, con- tradicting Haüy, he stated that not only did strontium carbonate crystallize in the same form as aragonite but also that lead carbon- ate took the same figure. These substances were isomorphous, and, similarly, calcite, magnesite, and iron carbonate were isomorphous. Calcite and aragonite, having the same chemical composition, dis- played what Mitscherlich termed the phenomenon of dimorphism. Accordingly, he revised his law in this statement: "An identical number of atoms, if combined in the same manner, will produce the same crystalline form, so that the crystal form does not depend upon the nature of the atoms, but upon their number and manner of combination."[51] Thus, the atoms of strontium carbonate and aragonite combined in the same manner so that the crystals of these two analogous chemical compounds displayed the phenomenon of isomorphism. But in calcite and aragonite, an identical number of atoms of the same elements arranged themselves so that the result- ing crystal forms and physical properties were completely differ- ent.

The more general term, polymorphism, was subsequently used to describe the phenomenon that Mitscherlich named dimorphism because certain substances crystallize in more than two forms. Mitscherlich recognized that additional study of the phenomenon of dimorphism was necessary because it had been proved to exist only in calcite and aragonite and postulated in the diamond and carbon. Actually, it had been observed in one other substance at least half a century before without any special note having been taken of it. In 1773 Antoine Baumé (1728-1804) had remarked that the crystals of sulfur obtained by allowing liquid sulfur to cool slowly were different from those produced by its sublimation.[52] Mitscherlich decided that the example of sulfur ought to suffice

---

[50] E. Mitscherlich, "Ueber das Verhältnis zwischen der chemischer Zusam- mensetzung und das Kristallform arseniksauerer und phosphorsauerer Salze," in A. Mitscherlich, *op. cit.*, pp. 133-173.

[51] *Ibid.*, p. 173.

[52] A. Baumé, *Chymie expérimentale et raisonée* (3 vols.; Paris, 1773), I, 241.

to prove his theory of dimorphism because sulfur was almost universally considered to be an element in which no vestige of any other element could be detected. In a memoir to the Academy of Berlin in 1822, he presented the results of an exhaustive examination of the orthorhombic and monoclinic modifications of sulfur. He believed these would remove the last trace of doubt concerning the validity of the hypothesis of dimorphism.[53] He was correct. On July 6, 1822, less than a month after Haüy's death, Gay-Lussac wrote a congratulatory letter in which he expressed his admiration and that of his colleagues for Mitscherlich's scientific achievement.[54]

There were attempts, of course, to determine why the phenomenon of polymorphism occurred, that is, why the atoms assumed different positions to produce such modifications of calcium carbonate as aragonite and calcite. Mitscherlich, for example, studied these latter substances in some detail. He noted that aragonite decomposed into calcite at higher temperatures but that calcite remained stable at different temperatures.[55] He observed that the crystals of sulfur exhibited a similar change upon heating. He pointed out that crystals of calcite and aragonite had been found in localities where it was obvious they had been formed from a liquid solution. On the other hand, both varieties of calcium carbonate had been discovered in places where geological conditions indicated they had crystallized from molten material. The evidence was confusing because the crusts of pieces of aragonite which had been covered with hot lava from Vesuvius had been transformed into calcite, whereas the cores had not been affected. But he admitted finally that any explanation of dimorphism was pure speculation in the contemporary state of science.

Nevertheless, the discovery of the phenomena of isomorphism and polymorphism dealt a severe blow to Haüy's concept of the specificity of the form of the integrant molecule. The figure of the molecule could be almost exactly the same if the number of chemically related atoms and their manner of aggregation were identical in two or more substances. Similarly, polymorphism demonstrated that the same chemical species could exist in at least

---

[53] E. Mitscherlich, "Ueber die Körper, welche in zwei verschiedenen Formen kristallisieren," *ADB* (1822-1823), pp. 43-48.
[54] A. Mitscherlich, *op. cit.*, p. 194n.
[55] E. Mitscherlich, "Ueber die Veränderung des Arragonits in Kalkspath," in *ibid.*, pp. 310-312.

two different forms that were unrelated by any laws of decrement. The form of the integrant molecule, then, was relegated to the position of being merely another characteristic, as it was in Werner's system, which should be taken into account along with physical and chemical properties as a means of identification.

Fig. 16. Calcite, CaCO₃, packing model. Ca, black; O, white; C, at the center of the CO₃ triangle, is not seen. The principal or c axis is vertical. The horizontal layers of Ca⁺⁺ ions alternate with horizontal layers of CO₃⁻⁻ ions. Source: Dana's Manual of Mineralogy (17th ed., © 1959), p. 331. By permission of the publisher, John Wiley & Sons, Inc.

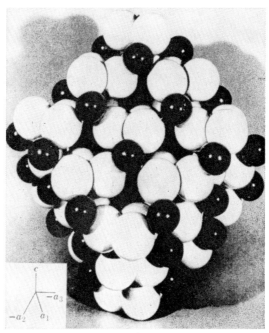

Simultaneously with these chemical studies, experimentation was taking place with respect to certain physical properties of crystalline matter. The results of this research also had the effect of modifying the basic concepts of Haüy's system. The study of the optical phenomena of double refraction played the most important part by far in contributing to the belief that crystals had unique axes in a physical sense, although the observation of pyroelectricity in certain crystals pointed to the same fact. This work simultaneously demonstrated that Haüy's method of arriving at the form of the integrant molecule by cleavage and clever guesswork was a dangerously fallible procedure.

Fig. 17. The arrangement of atoms in the Iceland crystal, CaCO₃, as determined by X-ray diffraction. Oxygen atoms are not shown in the top figure to avoid confusion. The arrangement of atoms in successive layers, perpendicular to the principal axis, EC, is shown below. The oxygen atoms are represented here by small black circles. The figures show only the relative positions of the atoms and do not indicate size. The CO₃ group is most clearly seen at r in the drawing of the middle layer. Note the symmetrical arrangement of the atoms around the principal axis. SOURCE: The Universe of Light, *by William Bragg, p. 184. Published by G. Bell & Sons, Ltd., London, 1933, and Dover Publications, Inc., New York, n.d.*

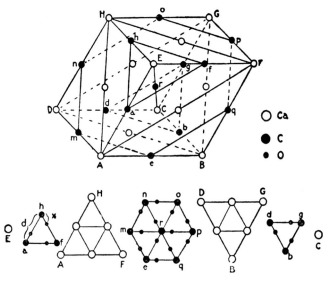

All crystals, except those belonging to the isometric system, display the phenomenon of the double refraction of light. In crystals belonging to the tetragonal and hexagonal systems, there is only one direction in which light may be transmitted without double refraction, that is, parallel to the vertical crystallographic axis (figs. 16, 17). A light ray entering the crystal and transmitted in any direction other than parallel to the vertical axis is divided into two rays that are polarized and vibrate at right angles to each

Fig. 18. Calcite, viewed normal to the rhombohedral face and showing double refraction. The double repetition of "calcite" at the top of the photograph is seen through a face cut on the specimen parallel to the base. SOURCE: Dana's Manual of Mineralogy *(17th ed., © 1959), p. 169. By permission of the publisher, John Wiley & Sons, Inc.*

other. One ray vibrates in a direction perpendicular to the direction of propagation and perpendicular to the vertical axis irrespective of the direction of that ray to the vertical axis. Because the path of the ray is predictable by the law of refraction, it is called the "ordinary" ray. The second ray vibrates in a plane containing it and the vertical axis; its direction of vibration is dependent upon the angle of inclination to the vertical axis. This is called the "extraordinary" ray because its path is contradictory to the law of refraction. Since there is only one unique direction of isotropy in these two systems, crystals belonging to them are termed optically uniaxial (figs. 18, 19). In a similar manner, owing to crystal structure, crystals belonging to the orthorhombic, monoclinic, and triclinic systems have two such directions in which no double refraction takes place; they are called optically biaxial. In

Fig. 19. Calcite, showing no double refraction. Viewed parallel to the principal or c axis. SOURCE: Dana's Manual of Mineralogy *(17th ed., © 1959), p. 169. By permission of the publisher, John Wiley & Sons, Inc.*

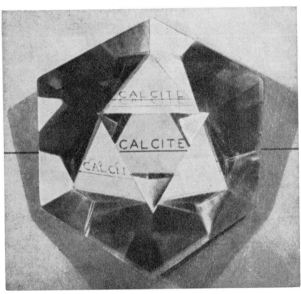

crystals for which the refractive index of the extraordinary ray is greater than that of the ordinary ray, the vibrations parallel to the vertical axis are those of the ray of least velocity, and the crystal is said to be optically positive. Crystals in which the vibrations parallel to the vertical axis, that is, the extraordinary ray, have the least refractive index are termed optically negative. In the majority of crystals, the amount of double refraction is small and difficult to detect with the naked eye. It was first noted in calcite, the Iceland crystal, by Erasmus Bartholinus.

In like manner, pyroelectricity is related to crystal structure. The simultaneous development of positive and negative charges of electricity at opposite ends of a crystallographic axis under certain conditions of temperature change is known as pyroelectricity. Pyroelectricity is permitted to crystals belonging to the twenty-one classes not possessing a center of symmetry, but actually does not occur in all of these because of certain other restrictions. One crystal displaying marked pyroelectricity is the tourmaline.

Franz Ulrich Aepinus (1724-1802), a professor of astronomy at Berlin, was the first to conduct a systematic investigation of the pyroelectricity effect of the tourmaline in 1756.[56] He remarked that the electrical properties of the tourmaline had long been known to jewelers. When tourmalines were heated on coals as a test of hardness, it was noticed that the gems alternately attracted and repulsed the cinders. Aepinus was interested principally in the change of polarity which the stone exhibited when it was heated. He supposed that it simultaneously possessed positive and negative electric fluids. When one of the sides of a heated tourmaline gave evidence of the presence of th positive fluid, the other side showed negative electrification. He found it was possible to change the polarity of the stone by heating one side and not the other, but he determined that it returned to its natural state when cooled. At first he believed that the positions showing positive and negative electricities depended upon the manner in which the stone was cut, but comparison with other tourmalines demonstrated that this belief was erroneous. He came to the conclusion that the cause must be sought in the interior structure of the gem.

Aepinus' experiments attracted the attention of Benjamin Wilson (1721-1788) who corroborated Aepinus' view that the direction of the fluid did not depend upon the figure of the gem but on some peculiar internal condition. In similar experiments, Wilson determined that the direction of the electric fluid moving through the crystal was always related to a particular direction: ". . . the direction of the electric fluid moving therein, *is always along the grain or shootings of the crystal.* . . . And the reason appears to be this, that the resistance the fluid meets with in passing through the gem, is less in that direction, than in any other."[57] Wilson and, almost simultaneously, John Canton (1718-1772) determined that

[56] F. Aepinus, "Mémoire concernant quelques nouvelles expériences électriques remarquables," *Histoire de l'Académie Royale des Sciences et Belles Lettres de Berlin*, XII (1756), 105.

[57] Benjamin Wilson, "Experiments on the tourmalin," *PT*, LI (1759), 308.

the pyroelectric effect depended upon changes in temperature rather than upon the fact that the gem was hotter than the ambient air.[58] As long as the temperature remained constant, the crystal was electrically neutral. During an increase in temperature, one ned became positively charged and the other negatively charged. During a decrease in temperature the polarities were reversed. Neither offered an explanation for this observation.

Haüy attempted to relate this phenomenon to the form of the tourmaline.[59] He determined that its primitive form was an obtuse rhombohedron, and he found that the centers of electrical action were only slightly removed from the extreme solid angles. He measured the electrical density at various points on the stone and ascertained that it diminished rapidly within short distances away from the poles and reached null points in positions situated at the middle of the stone. From the fact that both parts of a broken tourmaline exhibited pyroelectricity, he conjectured that each integrant molecule could be considered a small electric body, one extremity of which possessed positive and the other negative electricity. Although he did not relate the pyroelectric effect to any internal disposition of the elementary molecules, he did note that pyroelectric crystals departed from the symmetry of ordinary crystals in that the parts in which the two types of electricity resided, although similarly situated with respect to one another, differed in their configuration.[60] One part suffered decrements that were nonexistent on the opposite part or decrements took place on the opposite part which followed another law. In this he was referring to the fact that, when doubly terminated, tourmaline crystals usually show different forms at the opposite ends of a vertical axis.

The pyroelectric effect contributed, then, to a belief that crystals possessed unique axes in a physical sense. The study of the double refraction of light through crystals, however, lent the most powerful support to this view. Bartholinus had noted the fact that there was one position of the Iceland crystal through which two

---

[58] Benjamin Wilson, "Observations upon Some Gems Similar to the Tourmalin," *PT*, LII (1761), 443-447; John Canton, "Letter to Benjamin Franklin," *PT*, LIII (1762), 457.
[59] R. J. Haüy, "Mémoire sur les propriétés électriques de plusiers minéraux," *ADS* (1785), pp. 206-209; and "Extrait des observations sur la vertu électrique que plusiers minéraux acquièrent à l'aide de la chaleur," *JHN*, I (1792), 449-461.
[60] Haüy, *TDMı*, I, 237.

images of an object at the far side of the crystal could not be seen.[61] He dismissed the idea that the rhombohedral figure of the stone was the cause of the double refraction, after having attempted and failed to observe it in other crystals of a similar figure. He concluded that the cause must lie in some peculiarity in the body of the stone. Huygens, as we have shown, postulated that the Iceland crystal was composed of tiny ellipsoidal particles. According to his theory of the wave nature of light, the extraordinary refracted ray assumed an elliptical wave front while that of the ordinary ray remained spherical.[62] Thus, the directional structure of the crystal was of major importance in his theory, and double refraction, he remarked, did not occur in a direction parallel to the axis of the crystal because the two different propagations of light had an equal velocity in this direction. Elliptical light waves were not produced; spherical light waves remained unchanged.

Sir Isaac Newton rejected Huygens' theory that light consisted of waves and insisted, instead, that light was corpuscular in nature. He postulated that the light corpuscles had four sides or "quarters," two of which he termed sides of usual refraction and the other two, sides of unusual refraction. The unusual refraction of the Iceland crystal, he thought, was produced by a type of attraction possessed by certain sides of the rays and the particles of the crystal.

For if it were not for some kind of Disposition or Virtue lodged in certain sides both of the Rays, and of the Particles of the Crystal and not in their other Sides, . . . the Rays which fall perpendicularly on the Crystal would not be refracted towards that Coast rather than towards any other Coast.[63]

Newton believed that the presence of a position through which a double image of an object could not be seen supported his theory of double refraction, because in this direction "a Virtue or Disposition in those sides of the Rays . . . answers to, and sympathizes with that Virtue or Disposition of the Crystal."[64]

Haüy naturally was led to examine the phenomenon of double

[61] E. Bartholin, *Versuch mit dem doppeltbrechenden isländischen Kristall* (Leipzig, 1922), p. 14.
[62] Christiaan Huygens, *Treatise on Light*, trans. from French by Silvanus P. Thompson (Chicago, 1912), pp. 63-88.
[63] Sir Isaac Newton, *Opticks* (London, 1730: reprinted: New York, 1952), p. 373.
[64] *Ibid.*

refraction in his study of crystalline substances. He attempted always to refer the duplication of the images of objects seen through the crystal to the sides of the integrant molecules of such substances as calcite, rock crystal, sulfur, barite, and gypsum. Although he noted the fact that the absence of double refraction was related to an axial direction of the primitive forms of these substances, he did not notice the optical biaxiality of the last three substances. He did, however, remark that all substances whose primitive form was a cube, a regular octahedron, or a rhombic dodecahedron displayed only simple refraction. He accepted Newton's explanation of double refraction in its main outline, but he seemed reluctant to postulate any hypothesis that would relate the phenomenon to the internal structure of the integrant molecule.[65]

In a memoir published in 1807, Bernhardi emphasized that the theory of double refraction was connected intimately with that of crystallization.[66] Double refraction, he agreed, was evident only in irregularly crystallized bodies, and, although the property was found in different degrees in various substances, it appeared to follow the same rules in all. Two points or light poles, he stated, which were mathematically related to the structure of the crystal, appeared to be present in bodies that refracted double rays. Through these poles, a straight line or axis, termed the light axis, might be drawn. A beam of light incident on the crystal would, with two exceptions, be divided into two parts designated $+ L$ and $- L$, one of them following the ordinary law of refraction and the other being drawn closer to the light axis. The divergence of the two refracted rays depended upon the inclination and the position of the surfaces through which one looked and the direction of the incident ray. In only two instances was the refraction simple: when the crystal faces were parallel or perpendicular to the crystal axis. Working with calcite, quartz, emerald, and beryl, which are uniaxial, as well as zircon, vesuvianite, and mesotype (natrolite), which are biaxial, Bernhardi therefore actually must

[65] R. J. Haüy, "Sur la double réfraction du spath calcaire transparent," *JHN*, I (1792), 63, 158-160; "Sur la double réfraction du cristal de roche," *ibid.*, pp. 406-408; "Sur la double réfraction de plusiers substances minérales," *ADC*, XVII (1793), 140-156; and *TDM1*, I, 237.

[66] J. Bernhardi, "Beobachtung über die doppelte Strahlenbrechung einiger Körper, nebst einigen Gedanken über die allgemeine Theorie derselben," *GJ*, IV (1807), 230-258.

have seen the optical biaxiality of the latter, but he did not distinguish it from the uniaxiality of the former. But he did emphasize that double refraction was clearly dependent upon the material structure of the crystal and that there was an axis of refraction to which the phenomenon could be referred.

The determination of the actual connection between optical properties and crystalline form could commence, however, only after the discovery of the polarization of light by reflection in 1809 by Etienne Malus.[67] In Malus' view, the symmetrical disposition of the integrant molecules was the principal element of the phenomenon of double refraction. In ordinary refraction, the ratio between the sines of the angles of incidence and refraction was a constant quantity, but this relationship was variable in the case of the extraordinary ray. It depended not only upon the inclination of the ray incident upon the refracting surface but also upon its position relative to the axis of the integrant molecule of the crystal. The emergent ordinary and extraordinary rays had the same direction when the incident ray fell upon a face that was parallel or perpendicular to the axis of the crystal and emerged from a face parallel to the incident face. This circumstance, in Malus' opinion, gave a direct means of determining the axis of refraction of diverse crystalline substances. When the image viewed across two parallel faces was not doubled, the conclusion could be made that these faces were either parallel or perpendicular to the axis of the crystal. If one of these parallel faces was cut or ground to give another inclination and the image remained simple, then there was assurance that the first face was perpendicular to the axis of refraction. In the rhombohedron, the axis of refraction was the same as the crystal axis, but there was not sufficient information available, according to Malus, to state that this was the case a priori in other crystal forms.[68] Thus Malus, who was thoroughly familiar with Bernhardi's work, failed in the same way to distinguish uniaxial and biaxial crystals.

Assuming that the axis of double refraction and the crystal axis took the same direction in the primitive rhombohedron of calcite,

[67] E. Malus, "Sur une propriété de la lumière réfléchie," *Mémoires de Physique et de Chimie de la Société d'Arcueil*, II (1809), 143-158; and "Sur une propriété des forces repulsives qui agissent sur la lumière," *ibid.*, pp. 254-267.

[68] E. Malus, *Théorie de la double réfraction de la lumière dans les substances cristalisées* (Paris, 1810), pp. 95, 177.

Malus indicated that the value of the interfacial angles and that of the angle of the inclination of a face to the crystal axis which Haüy had assigned were slightly in error.[69] Also, according to Haüy, the most symmetrical form of aragonite, the right hexagonal prism, was an aggregate of four rhombohedral prisms, each of which had been formed by a law of decrement operating on the integrant molecule. The integrant molecule of aragonite was considered to be an irregular octahedron. Haüy's construction of the right hexagonal prism of aragonite assumed that each of the four crystal axes of the primitive octahedrons must be situated at different angles to one another. Malus determined first that the axis of double refraction of the primitive form given by Haüy did not correspond with its crystal axis, and, second, that the phenomenon of double refraction in aragonite was not compatible with Haüy's construction of the right hexagonal prism. Malus, however, did not challenge the selection of Haüy's primitive form nor the construction of the secondary form. He concluded that the axis of double refraction and the crystal axis did not correspond in irregular octahedral primitive forms.[70] Perhaps, had Malus experimented further, he would have found that aragonite was optically biaxial, but this discovery was reserved for Biot.

In 1812 David Brewster and Biot independently recognized that interference colors resulting from the passage of light through thin plates of doubly refracting crystals were weak indications of their double refracting property. Biot had experimented with a variety of mica (monoclinic), and Brewster with topaz (orthorhombic). Continuing their experiments, both Brewster and Biot arrived at the concept of optical biaxiality, but Biot was the first to publish his results.[71] Within a short time Biot determined also that uniaxial crystals could be divided into two classes. He assumed that the polarized light corpuscles in the extraordinary ray in quartz were attracted to the vertical axis whereas those in beryl were repelled from it, and he pointed out that the Iceland crystal had "beryl"

---

[69] *Ibid.*, p. 263.

[70] *Ibid.*, pp. 248-251.

[71] J. B. Biot, "Mémoire sur un nouveau genre d'oscillation que les molécules de la lumière éprouvent en traversent certains cristaux," *Mémoires de l'Institut* (1812), Part I, pp. 1-371; David Brewster, "On the Affections of Light Transmitted through Crystallized Bodies," *PT* (1812), Part I, pp. 187-218.

polarization.[72] Brewster, however, rejected, Biot's postulation of attractive and repulsive forces, relating the difference in the classes to the internal structure.

In crystals . . . the polarizing structure . . . must therefore depend on the form of their integrant molecules, and the variation in their density. . . . When these crystals have a spherical form diminishing in density towards an axis, and have these axes arranged by laws of crystallization, they will constitute a crystal of the positive class, . . . quartz. . . . When the density of the spheres increases towards their axes, their symmetrical combinations will constitute a crystal of the negative class, such as beryl.[73]

Hence, Biot's delineation of this difference as resulting from attractive and repulsive forces was replaced by the present description of such crystals as being optically positive and negative.

Both men, however, were quick to point out that the axes of double refraction corresponded to the crystal axes in many substances. According to Biot, the axis of double refraction proved that a crystal was an aggregate of material elements arranged in a symmetrical manner about the axis.[74] Asymmetry waas demonstrated by the fact that a light ray incident on the crystal in an oblique direction to this axis was doubly refracted. He emphasized that double refraction would be useful as an experimental index to analyze crystalline substances in an effort to determine the system of crystallization, particularly where the exterior form was not a sufficient indication of the interior crystalline state, for example, the mixed salts being studied by Beudant.

Over a period of years Brewster classified almost three hundred crystalline materials according to whether they had one, two, or three axes of double refraction, or, in modern terms, whether they were uniaxial, biaxial, or optically isotropic, respectively. He

[72] J. B. Biot, "Sur le découverte d'une propriété nouvelle dont jouissent les forces polarisent des certains cristaux," *Mémoires de l'Institut* (1812), Part II, pp. 19-30; and "Observations sur la nature des forces qui partagent les rayons lumineux dans les cristaux doués de la double réfraction," *Mémoires de l'Institut* (1813-1815), pp. 221-234.
[73] David Brewster, "On the Laws of Polarization and Double Refraction in Regularly Crystallized Bodies," *PT* (1818), p. 264.
[74] J. B. Biot, "Mémoire sur l'utilité des lois de la polarisation de la lumière, pour reconnaître l'état de cristallisation et de combinaison dans un grand nombre de cas où le système cristallin n'est pas immédiatement observable," *ADS* (1816), pp. 275-346.

pointed out that all crystals having the cube, the regular octahedron, or the rhombic dodecahedron as primitive forms have three axes of double refraction. Hence, in these crystals, no double refraction is evidenced. All crystals having the rhombohedron, the hexagonal prism, or the octahedron with isosceles triangular surfaces as primitive forms have one axis of double refraction. Substances with two axes of double refraction crystallize in other irregular forms. Hence, according to Brewster, the phenomenon of double refraction demonstrated the position of the crystal axes and could be used to determine the correct primitive forms of crystals. Brewster believed it was much more difficult to gain knowledge of the primitive form by cleavage and calculation than it was to test a crystal for double refraction. In Brewster's opinion, tungstate of lime provided a good example of the utility of this optical phenomenon in the derivation of the primitive form. Haüy believed that the form of the integrant molecule of this substance was a cube, but since it was doubly refracting, this figure was inadmissible and was accordingly changed. Brewster concluded:

The coincidence of the preceding deductions with the experimental results affords a strong presumption that we have been tracing the actual operations of nature. In the phenomena of crystals both with one and two axes, we have seen that these axes are coincident with some permanent line in the primitive or secondary form of the crystal.[75]

Within a short time, Brewster became bolder in his assertation of the connection of the crystalline form and the optical properties. He listed eleven minerals that could not have the primitive forms assigned to them by Haüy because their optical characteristics were not compatible with these figures. He announced in 1823 that eight of these substances had been assigned new forms in conformity with his observations. Boracite, for example, could not have a cubic integrant molecule because it was doubly refracting; it must have some other form. Brewster stated: ". . . optical phenomena are the necessary results of a mechanical structure and their indications must infallibly harmonize with the sound deductions of crystallography. . . . the optical mineralogist will only call a crystal cubic when it has no double refraction."[76]

---

[75] Brewster, "On the Laws of Polarization," p. 258.
[76] David Brewster, "Reply to Mr. Brooke's Observations on the Connexion between the Optical Structure of Minerals and Their Primitive Forms," *EPJ*, IX (1823), 361.

Further experimentation built up a mass of evidence in favor of the concept that crystals possessed unique axes of symmetry in a physical sense. In 1824 Mitscherlich published a memoir describing the effects of temperature change on the dimensions of a number of crystals of various substances.[77] First, he demonstrated that crystals that were regular geometric solids, like the cube, the regular octahedron, and the rhombic dodecahedron, expanded uniformly in all directions upon a temperature increase and evidenced no change in the value of their interfacial angles. Second, crystals having rhombohedra or hexagonal prisms for their primitive forms displayed unequal expansion in one direction when they were heated. For example, calcite expanded nonuniformly in the direction of its vertical axis, whereas its expansion was uniform in the other two axial directions. Hence, crystals with one axis of double refraction reacted in exactly the same manner to heat as to light. Third, all crystals shown to possess two axes of double refraction expanded unequally in all three directions when heated. Mitscherlich concluded from his experiments that the expansion of crystals upon heating was related to the axes of crystallization and that these axes corresponded with the optical axes. In 1830 Brewster stated confidently:

... it is obvious that the molecules of crystals have several axes of attraction, or lines along which they are most powerfully attracted, and in the direction of which they cohere with different degrees of force. ... the forces of double refraction are not resident in the molecules themselves, but are the immediate result of those mechanical forces by which these molecules constitute solid bodies.[78]

Haüy recognized the value of the phenomenon of double refraction in the identification of the integrant molecule of a substance, and he was willing to seek a new primitive form if this optical property indicated its absolute necessity. Further, he believed that, together with form, hardness, and specific weight, double refraction provided evidence of specificity. He pointed to this fact during the aragonite controversy. But he did not allow this directional property the importance Brewster attributed to it.

---

[77] E. Mitscherlich, "Ueber das Verhältnis der Form der kristalliserten Körper zur Ausdehnung die Wärme," in A. Mitscherlich, op. cit., p. 195.

[78] David Brewster, "On the Production of Regular Double Refraction in the Molecules of Bodies by Simple Pressure; with Observations on the Origin of the Doubly Refracting Structure," PT (1830), pp. 87-95.

This fact is evident in that, to the end of his life, Haüy believed the angle of inclination of the cleavage face of the primitive calcite rhombohedron to the vertical axis was exactly 45°, whereas many optical tests had, by that time, demonstrated that this conclusion was erroneous. Thus, the experimental results of physicists had a similar effect on Haüy's method and system as those of the chemists. As the phenomena of isomorphism and dimorphism had proved that the form of the integrant molecule could not be considered in itself a specific characteristic of a chemical compound, optical properties demonstrated that the primitive form and the integrant molecule could not dependably be derived by cleavage of the secondary forms. Haüy's system had been shown to be in error in many respects, because if the figure of the integrant molecule was different from that which had been ascribed to it by Haüy, it followed necessarily that the laws of decrement assigned for the production of the secondary forms were incorrect also. Obviously, either Haüy's system must be entirely revised by the determination of the correct primitive forms, or another system of crystal analysis should be adopted. The latter course prevailed, but, in essence, the new methods, using Haüy's laws of decrement as a basis, were extensions of these discoveries. But, also, the new systems included a novel view of crystal symmetry and were to lead, in time, to the establishment of modern crystal classifications.

CHAPTER VI

The Concept
of Crystal Symmetry

The degree of external symmetry possessed by any crystalline substance is determined by the number and type of symmetry operations which can be performed upon it. The rotation of a crystal around a crystallographic axis a certain number of degrees so as to bring it into coincidence with its original aspect constitutes one type of symmetry operation which defines the symmetry of a crystal. It has the symmetry of reflection across a plane if one half of the crystal is seen to be the mirror image of the other half. Further, the crystal has the symmetry of inversion through a center if a point on its surface has a corresponding point on the opposite side equidistant from the center along an imaginary line drawn through the center and the two points. Considered in this manner, any crystal will correspond to the symmetry relations of one of thirty-two possible crystal classes or point groups, which can in turn be grouped into six crystal systems. For example, the cube has a center about which it is symmetrical. It has, also, three axes of four-fold rotation through the centers of its faces; a rotation of the cube 90 degrees around any of these axes will bring the cube into coincidence. Similarly, it has four axes of three-fold symmetry through its diagonals, and six axes of two-fold symmetry, each of which bisects two opposite edges of the cube. Further, it has nine planes of symmetry, three through the centers of its faces and six through opposed edges. All these symmetry elements together give the highest possible degree of crystallographic symmetry and define the hexoctahedral class of the isometric system, of which the regular octahedron and the rhombic dodecahedron are members along with the cube.

Not only do the crystal classes and systems serve to define the degree of external symmetry pertaining to any crystalline substance, but they also distinguish certain vectorial properties of crystals. Double refraction of light is displayed by crystals of all systems except the isometric, but whereas crystals belonging to the tetragonal and hexagonal systems are optically uniaxial, those of the orthorhombic, monoclinic, and triclinic systems are optically biaxial. Only those substances crystallizing in the twenty-one crystal classes that have no center of symmetry are permitted the property of piezoelectricity, wherein an electric charge is developed on the surface of a crystal when pressure is exerted at the ends of a crystallographic axis. In like manner, pyroelectricity is restricted to certain crystal classes.

The internal symmetry of a crystal, on the other hand, is viewed with reference to the orderly arrangement of the particles that compose it, although, of course, the external symmetry reflects this order. The crystal is pictured as being constructed of a large number of minute units which are repeated in three dimensions. Atoms, ions, ionic groups, or molecules are arranged on points or nodes of a three-dimensional space lattice in such a manner that they all have identical environments, and the lattice is defined by the three directions and by the distances along these directions at which the pattern, thus formed, is repeated. In the nineteenth century, it was demonstrated from geometrical considerations that it was possible to have only fourteen varieties of space lattices, since other arrangements of points destroyed the requirement that the environment about every lattice point be identical. Later, it was seen that these space lattices, in addition to translation in space, might be subject to two other symmetry operations: rotation in a screwlike manner and also what is termed "glide reflection." Within a short time it was determined that a total of 230 space groups were mathematically possible by combining all the symmetry operations in all possible ways.

A full consideration of the historical development of the concept of the space lattice is not undertaken in this book, although there is some discussion of early ideas of internal crystal symmetry at the end of this chapter. Rather, we are here engaged in describing the progress of thought concerned with the external symmetry of crystals. In particular, it will be seen how this line of development resulted primarily from the consideration of the structure of matter from the polar rather than from the molecular point of view. It was primarily with this basic presconception—that the underlying structure of matter should be considered in terms of dynamic forces of attraction and repulsion inhering in poles—that Christian Samuel Weiss postulated a radically new conception of crystal symmetry and systems of crystal classification, and, in the process, restated Haüy's laws of decrement in such a way that they took the modern form of the law of rational intercepts. Taken as a whole, Weiss's work constituted a major revision of Haüy's method and provided a firm basis for the later nineteenth-century development of mathematical crystallography.

In chapter ii I pointed out that the polar concept of the structure of matter had been held by Leibniz, Swedenborg, and Boscovich, and that Kant was also an adherent of this theory. Kant believed that the dynamic philosophy of matter, as he termed it, was far more suitable and more advantageous to experimental philosophy than the atomistic view. Matter, in his opinion, was divisible to infinity into parts, each of which was again matter. By postulating an inherent repulsive force, the fact of the impenetrability of matter might be established, but with only repulsive forces acting, all spaces, he stated, would be empty, so that an attractive force must also be assumed. Macroscopic matter, then, could be viewed as a resultant of the action of these forces, and science was led directly toward the goal of the determination of the proper moving forces of matter and to the laws governing these forces. This concept limited the freedom of assuming intermediate spaces and fundamental bodies of a definite figure which were basic hypotheses of the atomistic or molecular theory of matter. Neither of these hypotheses, Kant stated, could be defined satisfactorily, nor could the physical entities that were postulated be discovered by any experimental techniques.[1]

The main direction of Kant's argument was toward economy; by the reduction of assumptions, one moved toward simplicity. He did not, as Boscovich had done, attempt to explain in any detail how the varieties of matters arose from the postulated primitive forces of attraction and repulsion; how, for example, a solid was

---

[1] Immanuel Kant, *The Metaphysical Foundations of Natural Science*, trans. from German by Ernest Belford Bax (London, 1883), pp. 172-185, 210-211.

formed from a fluid. He left the solution of these problems to others.

The *Naturphilosophie* movement in Germany in the late eighteenth and early nineteenth centuries incorporated the polar concept of matter among its tenets. Friedrich Schelling, one of the foremost proponents of this philosophy, recoiled from the abstract, static, mechanistic trend of science in his time, believing in a pantheistic manner that single phenomena incorporated or represented an ideal content that was derived from the Absolute. There was, he believed, a universal formative power in nature which produced individual things with assigned roles, and it should be the task of science to determine by what interactions these phenomena did execute or fulfill this purpose. Formative processes caught Schelling's attention, and he believed that it was an excuse for ignorance, merely because chemical processes were in his time better known than animal processes, to call vegetation and life chemical processes, rather than to call many chemical processes incomplete organization processes. Thus the formative process or inclination of nature to crystallization was one Schelling viewed as being clarified or perfected in vegetative or animal formations.

Structure and form, Schelling taught, were implicit in the idea of individuality. In all life everything that either itself or by the hand of man received a certain figure should be considered an individual. Each solid body, therefore, possessed a kind of individuality, such that each change from a fluid to a solid state was connected with a crystallization, that is, formation with a certain figure. In the continuity of the fluid condition, he stated, no part could be distinguished from another by its figure. But, upon solidification, the cohesion of form and matter was revealed in the many products of inorganic nature which crystallized in forms proper to them, if the conditions of crystallization were quiet and undisturbed. One should look for the reason for this in nothing other than what he termed the primitive quality (*ursprünglichen Qualität*), and, since the positive principle of all crystallizations was without doubt the same, one should search for a negative principle to account for different crystallizations.[2] Thus, he was critical of Haüy:

[2] Friedrich von Schelling, *Von der Weltseele* (Hamburg, 1798), pp. 189, 219.

All crystallizations [with Haüy] are to be regarded as secondary formations, which arise from the various aggregations of unalterable primitive forms, because such a derivation does allow mathematical construction; yet, this is merely a clever game, because it can in no way be proved that so simple a form is itself not secondary.[3]

Schelling mused on the connection of the polar forces that were apparently active in nature and seemed to manifest themselves in chemical combination, electricity, and magnetism, with Schelling himself seeing Sir Humphry Davy's electrochemical theory and Oersted's and Faraday's discoveries of electromagnetic phenomena as vindications of his earlier ideas.

From April, 1796, to August, 1798, Schelling was in Leipzig, where he studied mathematics, physics, and medicine. It was there that his *Ideen zur Philosophie der Natur* appeared at Easter, 1797, and *Von der Weltseele* a year later, and it was in this period that Christian Samuel Weiss, himself a medical student at the university, met Schelling and fell under the influence of his ideas of nature. Weiss, a brilliant student who appeared equally at home in all the sciences, received his doctorate from Leipzig in 1800 at the age of twenty.[4] In the following year he went to Berlin where he attended the mineralogical lectures of Dietrich Karsten, the counselor of mines of Prussia. He also busied himself in the laboratory of Martin Klaproth, who was laying the foundations for the quantitative analysis of minerals, and he came into contact with the great geologist, Leopold von Buch (1774-1852), who was at this time gathering data that would aid the eclipse of Werner's Neptunist theory. In 1802 and 1803 Weiss studied mineralogy under Werner at Freiberg and during this period prepared, under Karsten's direction, a German translation of Haüy's *Traité de mineralogie*, the first volume of which appeared in 1804. At the end there was included a lengthy article by Weiss entitled "Dynamische Ansicht

[3] *Ibid.*, p. 227.
[4] The biographical information on Weiss was obtained primarily from Martin Websky *et al., Gedenkworte am Tage der Feier des hundertjährigen Geburtstages von Christian Samuel Weiss den 3 März 1880* (n.p., [1880?]). See also Emil Fischer, "Christian Samuel Weiss und seine Bedeutung für die Entwicklung der Krystallographie," *Wissenschaftliche Zeitschrift der Humboldt-Universität zu Berlin*, XI (1962), 249-255; and "Christian Samuel Weiss und die zeitgenössische Philosophie (Fichte, Schelling)," *Forschungen und Fortschritte*, XXXVII (1963), 141-143.

der Kristallisation." In it Weiss outlined in detail how crystallization could be understood by reference to the polar theory of
matter, at the same time showing his dependence in this regard
upon the thought of Kant and Schelling.

Following them, Weiss supposed that there was in nature not
only an attractive but also a repulsive force.[5] The whole of nature
was a development of the action of these two forces. One could not
dare attribute an absolute existence to matter. There were only
poles that owed their existence to the fact that they were separated by repulsive forces from other poles. If the poles were to
coalesce completely, matter would vanish. The true idea of chemical combination, then, was that two or more poles interpenetrated
one another only partially and thus occupied space. The fundamental process of nature as creation from nothing, a development
from zero, was the operation of attractive forces which caused
poles to unite and the operation of repulsive forces which prevented their complete fusion and consequent obliteration. The
forces were inherent in the poles; they were not impressed external forces.

Either the attractive or the repulsive force became dominant
after the union of the poles, Weiss explained, although conflict
between the two continued. The initial union of poles which produced the matter of oxygen, for example, resulted in the excess
of an attractive force, and, similarly, hydrogen possessed an excess
of repulsive force. The attractive and repulsive forces of the polar
unions were in sensible equilibrium in fluids, but if the repulsive
forces gained in intensity, crystallization occurred. During the
crystallization, however, the repulsive forces were restrained in
varying degrees, and this condition caused variations in the crystal
form. The formation of ice from oxygen and hydrogen was ac-
companied by the creation of repulsive forces in the solid which
had a directional character and were therefore responsible for its
characteristic form. Owing to the directional nature of the repulsive forces inherent in various crystallizing matters, faces were
constructed which formed specific angles with one another. The
manner of solidification, then, depended upon the resultant direc-

_____

[5] R. Chenevix, "Observations sur une mémoire du docteur Christian Samuel Weiss, imprimé dans la Minéralogie de M. Haüy, traduite en allemand
par Dietrich-Ludwig-Gustave Karsten, conseiller des mines de S. M. Prussienne, etc., avec quelques remarques sur cette traduction," *ADC*, LII (1804),
308-339.

tion of the repulsive forces created by the union of the forces of the primary matters. Secondary directions of crystallization were derived from the primitive directions, so that secondary faces were lawfully inclined at certain definite angles. Cleavage could be explained by reference to the direction in which the repulsive forces operated.

The French reviewer of Weiss's article, Richard Chenevix (1774-1830), an expatriate Irishman living in France, could not have been more scornful. He apologized for misuse of the French language which was necessitated in the translation of the prolix Germanic utterances. He stated that Weiss believed his theory was clear and profound whereas, in fact, it was an unintelligible chimera. It was disavowed by reason, he asserted, and could not be considered as a scientific achievement in any sense. Despite Chenevix's complete rejection of Weiss's ideas, however, it is apparent that Weiss had proposed some important revisions with respect to crystal theory. He did not mention primitive forms in the sense that they were used by Haüy because, with Schelling, he believed that the primitive form itself was secondary. With the elimination of integrant molecules, there was no need for the supposition of grooves or trenches, however small, on the crystal faces. Weiss's construct yielded primary and secondary crystals which had true faces. But, more important, Weiss had clearly indicated the vectorial nature of crystallization. His theory emphasized that direction was a characteristic of a crystal type. The science of crystallography had to take direction into account, and from the consideration of direction, Weiss developed the notion of axes of symmetry within the crystal.

In 1805 Weiss commenced a three-year journey through southern Germany, Austria, Switzerland, and France, during which time he became convinced with von Buch that Werner's geological theory was erroneous. He visited his old friend Schelling at Munich and went to Paris for two extended stays from April to July, 1807, and from December, 1807, to August, 1808. Here he was intimate with the French scientists Cuvier, Brongniart, Ampère, and, of course, with Haüy and Haüy's assistant, Brochant de Villiers. Weiss complained about the despotic demeanor of the leading French scientists. They demanded, he reported, that their students and followers be completely acquiescent to the master's theoretical pronouncements. Haüy was certainly one of those to whom Weiss referred; Haüy castigated de Bournon in print for not giving sufficient consideration to his authority as a savant with a reputation.[6] It appears that Haüy was, for a while, able to overlook Weiss's self-confidence and brashness in view of his brilliance, but this situation could not last because fundamental differences existed in their respective theories of the structure of crystalline matter. In June, 1808, Haüy terminated the working relationship that had been established, telling Weiss that he had lost his reputation when Weiss refused to yield to Haüy's views. Weiss wrote: "Previously, I have stood uncommonly well with him [Haüy].... he has enough proof how much and where I have attempted to recognize his merit, but he is not satisfied. Tyranni- cally . . . he wishes more, he wishes everything; and there my neck is unbowed."[7] Although a position in physics had been vacant in Leipzig for some time, Weiss was not called to it until August, 1808, because of reservations concerning his commitment to the *Naturphilosophie* of Schelling's type. Upon receiving this post, Weiss read his dissertation, *De Indagando formarum cristallinarum caractere geometrico principali*, which outlined the basic differ- ences that existed between Haüy and himself and was the initial step in the development of his ideas of external crystal symmetry.

The idea of an axis around which the crystal was symmetrical was, of course, not new. Huygens used the concept in his crys- tallographic analysis of the Iceland crystal, and all Haüy's studies assumed the existence of an axis through the primitive form. Haüy, however, had used the axis principally as a reference direction from which he derived the limit ratios of the diagonals of the faces of the primitive forms. These ratios were the most impor- tant aspects of the primitive forms or integrant molecules. There was no clear-cut concept of symmetry with respect to more than one crystallographic axis. A review of Haüy's *Traité de miner- alogie*, for example, included the statement:

It is an important remark, that whatever may be the variety of form, the forms in complete crystals will always be symmetrical. There are two sorts of symmetry, the perfect and the imperfect. In the perfect the right is symmetrical with the left and the top with the bottom, but in the imperfect, the top is not symmetrical with the bottom.[8]

[6] R. J. Haüy, "Mémoire sur une loi de cristallisation, appelée *loi de symé- trie*," *Mémoires du Museum d'Histoire Naturelle*, I (1815), 219.

[7] Websky *et al.*, *op. cit.*, p. xi.

[8] A. Q. Buée, "Outlines of the Mineralogical Systems of Romé de l'Isle and the Abbé Haüy: with Observations," *Nicholson's Journal*, IX (1804), 82.

To Haüy and his followers symmetry consisted primarily in the uniform operation of the laws of decrement at edges and angles of the primitive form. De Bournon, for example, stated that it was a general rule that suffered few exceptions, that when a primitive form was symmetrical at its edges and angles, as in a cube, the regular decrement that took place at one of the edges would occur at other edges that were similarly placed. He pointed out that an exception occurred in the derivation of the hexagonal prism of calcite. Decrement operated only on the obtuse facial lateral solid angles without occurring on the acute angles.[9]

In 1815 Haüy published a series of memoirs in which he stated that the laws of decrement were subordinate to another law, which he called the law of symmetry.[10] He believed this law was characterized by its generality and uniformity and consisted in the fact that the same type of decrement operated on all parts of the primitive form which were identically positioned. These identical parts could be identified by the fact that when the position of the nucleus was changed with respect to the eye, the symmetrical elements always presented themselves in the same manner. Here he postulated the idea of the symmetry operation of axial rotation, but he did not develop it. He took each primitive form and by an example showed that certain secondary forms followed necessarily from it by the operation of the laws of decrement. But he admitted that there were exceptions to the law. *Cobalt gris* (cobaltine) was one example. Not recognizing the existence of a horizontal axis of two-fold symmetry, Haüy could not include this mineral in his scheme. The decrements that acted on the cube which was the primitive form, he stated, operated only on some of the edges and angles. But he argued that this case was very rare and depended upon some peculiar circumstance, the cause of which was then unknown. As for the tourmaline or other substances that became electrified by heat, he attributed the difference in configuration of the parts of these crystals which contained the electric poles to a cause "which diverts crystallization from the course that it would follow, if it were left to itself."[11]

Weiss's ideas were in direct contrast with this approach. In his dissertation he did not dwell on an exposition of the nature of the

---

[9] J. L. de Bournon, *Traité complet de la chaux carbonatée et de l'arragonite* (3 vols.; London, 1808), II, 213.

[10] Haüy, *op. cit.*, pp. 81-101, 206-225, 273-298, 341-352.

[11] *Ibid.*, p. 87n.

forces which caused crystallization, as he had in his first article.[12] He mentioned merely his view that crystal forms were the necessary results of generating forces, whose direction determined the form. He was concerned primarily with delineating the concept of crystallographic axes and the relationship that the elements of a crystal had with them. All crystals, Weiss asserted, had an axis that was a unique, principal, and dominating direction. Such crystal forms as the cube had not one axis but three. This condition gave the cube a higher degree of regularity. The cube should be considered a unique variety of the rhombohedron whose faces could be related to three crystallographic axes as well. These axes were in no sense to be considered merely geometrically. They had definite physical significance. The path of light aberration in the phenomenon of double refraction, for example, was related to the angle that the incident ray made with the crystallographic axis and could serve to determine the position of the axis in the crystal form. In determining the primitive form of calcite, Weiss stated, Haüy had considered the relationship between the two diagonals of a face as the basic element of the crystal form. This fundamental relation was, in Haüy's view, the principal geometric characteristic of the crystal. To Weiss, Haüy had erred, because this relation was not fundamental. The principal geometric characteristic should be founded on elements that had a direct relation with the crystallographic axis. The diagonals of a face did not have this direct relation, but the relation between the sines and cosines of the angles of the inclination of a face to the axis did. Every face of a crystal had this direct relation with the crystallographic axis. Hence, the crystal form could be analyzed and described in terms of these direct relations. Further, because of his preoccupation with crystallographic axes that he believed had physical reality, Weiss insisted that secondary forms could be determined without the necessity of employing the notion of Haüy's fictional subtractive molecules.

Brochant de Villiers, who translated Weiss's dissertation into French, pointed out that the new relationships which Weiss maintained were basic owed their exactness to the work of Haüy. He was correct. It is true that at this stage Weiss did work back and

---

[12] "Dissertatio de Indagando formarum cristallinarum caractere geometrico principali, ou Mémoire sur la détermination du caractère geometrique principal des formes cristallines," trans. from Latin by Brochant de Villiers, *Journal des Mines*, XIX (1811), 349-391, 401-444.

derive the relations of the crystal faces to the axis from Haüy's published values, but this condition did not last. Weiss had set the stage for the geometrical treatment of crystal forms which made it possible to express the basic law of the rationality of the indices in its correct form.

The main development of Weiss's thought occurred between 1811 and 1815. In December, 1815, he read a memoir to the Academy of Berlin in which many of the modern aspects of the science of crystallography were presented for the first time.[13] He classified all crystals into two main divisions: those that had three axes at right angles to one another, and those that had four axes, three of which were equally separated and perpendicular to the fourth. In the first division he considered three possibilities: the lengths of the three axes could be equal; the lengths of two axes could be equal and different in length from that of the third axis; and the lengths of all three axes could be different. Thus, Weiss distinguished the present isometric, tetragonal, orthorhombic, and hexagonal crystal systems. His term for the isometric system was "spherohedral." This nomenclature raises an important point. Weiss had begun to employ spherical trigonometry to analyze crystal structures. The importance of the use of this technique was pointed out a few years later by Mitscherlich:

Haüy and his students used only plane trigonometry; it is, however, applicable only if the dimensions of the axes of the primitive form stand in simple relationships to one another, which indeed, in the class of primitive forms that Haüy called "limit" forms, is true. With all other primitive forms that have been determined by exact instruments, this simple relationship has not been found. The measurements of Biot, Malus, Wollaston, and Phillips have proven that the assumption of a simple relationship of dimensions, on which the method of Haüy is practical, is involved with great difficulties and not based on facts.[14]

Weiss stated that the principal forms belonging to the spherohedral system were the cube, the regular octahedron, and the rhombic dodecahedron. Thus, he recognized that three of Haüy's six primitive forms were characterized by the same kind of symmetry. Configurations derived from these forms were the trape-

---

[13] C. S. Weiss, "Uebersichliche Darstellung des verschiedenen natürlichen Abteilungen der Kristallisations-systeme," *ADB* (1814-1815), pp. 289-344.

[14] A. Mitscherlich, ed., *Gesammelte Schriften von Eilhard Mitscherlich* (Berlin, 1896), p. 137n.

zohedron with twenty-four trapezoidal faces, the trisoctahedron with twenty-four triangular faces, and the hexoctahedron with forty-eight triangular faces. The three equal axes at right angles to one another created eight octants. Based on symmetry considerations, the maximum possible number of similar faces at different inclinations in each octant was six. Hence, Weiss stated that the maximum possible number of similar plane surfaces in the spherohedral system was forty-eight. He saw, however, that further subdivision of the spherohedral system was necessary. The homospherohedral class should include those forms in which all similarly shaped surfaces were fully present. Hence, Weiss distinguished the present hexoctahedral or holohedral class of the isometric system. As examples of this class, Weiss correctly listed diamond, fluorspar, spinel, and garnet. When all the similarly shaped faces were not present, the forms fell into one of the two subdivisions of the spherohedral system, which Weiss called the tetrahedral hemispherohedral class and the pentagonal hemispherohedral class, corresponding to the present hextetrahedral and diploidal classes of the isometric system. Again, he correctly identified sphalerite as an example of the former and pyrite as an example of the latter class.

Weiss gave the name "four-membered" (*vierglieder*) to the present tetragonal system, characterized by three mutually perpendicular axes, only two of which are equal in length. He demonstrated that the crystal forms of this system could have a maximum of sixteen similar surfaces. The state of development of Weiss's ideas, however, took into account only axial symmetry operations, so that he was unable to distinguish any subdivisions of this or any other crystal system which introduced other symmetry operations. In the four-membered system, zircon, zinc vitriol, and rutile were listed correctly as examples.

Weiss divided the crystal system that had three unequal mutually perpendicular axes into three separate classes. One of these classes, the "two-and-two-membered" (*zwei-und-zwei-glieder*), corresponds to the present orthorhombic crystal system. He recognized that crystals belonging to this system could have a maximum number of eight similar surfaces and listed topaz, niter, and aragonite as examples. But he committed a major error in his treatment of the other two crystal classes, which would otherwise correspond to the present monoclinic and triclinic systems. He retained mutually perpendicular axes for these two subdivisions. He assumed, for example, that if the primitive form of a monoclinic crystal was a prism with the base inclined obliquely to the vertical axis, a perpendicular from the axial center would yield a rational division of the oblique base line. Nevertheless, he was able to classify feldspar, gadolinite, and borax into the "two-and-one-membered" or monoclinic system and copper vitriol (chalcanthite) into the "one-and-one-membered" or triclinic system. In treating the divisions corresponding to the present hexagonal system, Weiss distinguished the hexagonal or "six-membered" subsystem and the rhombohedral or "three-and-three-membered" subsystem, characterized by sixfold and threefold symmetry axes, respectively. He correctly placed calcite, cinnabar, and tourmaline in the latter and beryl and apatite in the former subsystems. Weiss did err in this first attempt, however, by referring several minerals to the wrong systems.

It should be emphasized that Weiss did not believe the systems or crystal subdivisions he proposed were mere geometrical constructs. As the title of his memoir indicated, these were natural divisions. The variables that characterized each division and subdivision depended upon the operation of natural processes. The forces producing crystals could be treated mathematically by reference to the configuration of the end products of their action. Physical natural laws operated to allow the production of these classes of crystals only. Weiss did recognize, however, that further divisions might be possible within his systems, and he believed that studies should be made with this goal in mind. He appeared to weigh the possibility of the existence of crystal systems that did not have mutually perpendicular axes. He stated that closer research into the problem of the analysis of crystal forms ought to determine whether all crystal systems, with the exception of the hexagonal, were to be based on their reference to mutually perpendicular axes. It should be noted, also, that Weiss's method abandoned the concept of primitive forms. He referred Haüy's forms to the crystal systems or subsystems to which they belonged.

Weiss's revision of the science of crystallography was well received in Germany. It had the effect, however, of giving crystallography the character of a strictly mathematical science, which Haüy's method had not fully accomplished. Hitherto crystallography had been considered either an important adjunct to or a handmaiden of the sciences of chemistry and mineralogy. With

Haüy the value of crystallography lay in the fact that the determination of the primitive form and integrant molecule of a substance led directly to the definition of a species. With such mineralogists as Werner, the study of the external forms of crystals was important as a useful tool for the purpose of identification. The introduction of Weiss's systematic approach caused a gradual change in these attitudes. Henceforth, the chemist and mineralogist had to refer either primitive or secondary crystal forms to mathematically defined classes founded on considerations of symmetry. With the work of Berzelius, the species came to be defined in purely chemical terminology. The crystallographer became concerned with establishing the mathematical relationships evidenced in the crystalline end products of natural processes. The new attitude was demonstrated in an announcement by the mathematics class of the Berlin Academy of an open competition for the year 1817:

First, give a precise description of any crystallization whatsoever, either of calcite, barite, fluorspar, or a salt, or whatever one wishes, not in the technical language of the mineralogist, which is foreign to most mathematicians, but in clear geometrical terms, and, particularly, do not define the faces of the nucleus hypothetically, but in accordance with positive observation. Second, find a hypothesis concerning the laws of attraction by reference to which the internal structure of the crystal can be explained in accordance with the precepts of mechanics and represented by analytical formulas.[15]

This statement presented two clear objectives for research in the science of crystallography: Attempts should be made to explain mathematically the directional character of the bonding forces inherent in the crystallization process, and a clear, concise mathematical language should be invented in order to describe the resulting crystal form.

It was Weiss who accomplished this latter task.[16] He admitted that Haüy's system of notation was satisfactory with most simple crystal forms where the laws of decrement which operated on the nucleus were not complicated. But he said there were many cases in which Haüy's designation presented a double meaning, a con-

15 *ADB* (1814-1815), p. 5.
16 C. S. Weiss, "Ueber ein verbesserte Methods für die Bezeichnung der verschiedenen Flächen eines Kristallisations-systeme, nebst Bermerkungen über den Zustand von Polarisirung der Seiten in den Linien der kristallinischen Struktur," *ADB* (1816-1817), pp. 286-314.

dition that led to confusion and misunderstanding. He pointed out that in some instances the operation of the same intermediate law of decrement could result in three completely different designations for the same surface. The fact that they did describe the same face was in no way evident; calculations had to be made in order to determine that they were. The reason for this anomaly lay in the fact that the new surface could be regarded as having proceeded from various intersecting surfaces of the primitive form. It was fully optional from which of the intersecting surfaces one could regard the decrement as operating. For example, in respect to quartz, the various ways one could regard the operation of the laws of decrement in Haüy's notation produced these expressions for the same secondary surface. $(E^2B^1D^2)$, $(\frac{1}{4}EB^2D^1)$, or $(\frac{1}{2}ED^1D^4)$. Weiss claimed that any student would believe at first glance that these three notations indicated three different faces. Intense concentration and calculation were necessary to determine the fact that they did not.

Furthermore, according to Weiss, Haüy's notations complicated the calculation of the interfacial angles because these symbols showed only the relationship of the secondary surface to the primitive form. In every case it was necessary to refer to the angles of the primitive form and study the effect of each decrement mathematically in order to arrive at the interfacial angles of the new form. But these values were the essentials being sought. Weiss thought that an improved notation should allow the immediate calculation of the interfacial angles from the designation of the secondary face. He believed that this objective could be accomplished by referring the inclination of any face whatsoever to the three mutually perpendicular axes of the one main division of his crystal system or to the four axes of the other division. He designated the two mutually perpendicular horizontal axes a and b and the vertical axis c. He explained that the lengths a face intercepted on these axes respectively not only determined the shape of the crystal by axial symmetry operations but also resulted in the ability to calculate the interfacial angles directly. They were the essential keys in his notations. For example, if the face $\boxed{a :b :c}$ was considered, three unequal axial intercepts would describe the surface of an octahedron with scalene triangular faces (orthorhombic dipyramid). If two of the axial intercepts were equal and different from the third, the face would be that of an octahedron with isosceles triangular faces (tetragonal dipyramid). This face would be described more properly by the notation $\boxed{\text{a} :\text{a} :\text{c}}$. The designation $\boxed{\text{a} :\text{a} :\text{a}}$ should be employed if all three axial intercepts were equal, and the face would be that of a regular octahedron. The notation $\boxed{\text{a} :\text{b} :2\text{c}}$ would indicate immediately that the vertical axial intercept was double that of the preceding instance. An infinity symbol was used in front of the axial letter to designate a surface parallel to that axis, so the notation $\boxed{\text{a} :\text{b} : \infty \text{ c}}$ denoted the side surfaces of a quadrilateral prism. Weiss pointed out that determination of the interfacial angle was relatively easy by the concurrent use of this notational system and the sine-cosine relationship. In the case of the face $\boxed{\text{a} :\text{b} :2\text{c}}$, for example, the value of the sine of the angle of inclination of the face to the axis c was to its cosine as $ab \sqrt{a^2 - b^2} : 2c$. The reciprocal of this proportion gave the value of half the angle this face made with the corresponding face below the horizontal axes. Weiss gave a similar notation to his crystal division that corresponded to the present hexagonal system. For example, the designation

$$\boxed{\begin{array}{c} \infty \text{ c} \\ \hline \text{a} :\text{a} : \infty \text{ a} \end{array}}$$

was the symbol for a surface that formed the side of a hexagonal prism. At first, Weiss suggested merely that his symbols be used in conjunction with Haüy's designations, since he recognized that the latter were universally known and their use was strongly entrenched. But the clarity and simplicity of Weiss's notations led to their swift adoption by German scientists. The present Miller indexes, which are the reciprocals of Weiss's symbols, were introduced later. This notational system was, of course, an improvement, for it eliminated the necessity for fractions and infinity signs.

During the period when Weiss was revising Haüy's method and notational system, Friedrich Mohs (1773?-1839), Werner's successor at Freiberg, was also studying the possibility of the establishment of crystal systems based on axial symmetry. In 1822 he published a mineralogical treatise in which he classified crystals into systems in much the same manner as had Weiss. The fact that he did not mention Weiss's name in this work caused a sharp interchange between the two men on the question of priority.[17] Weiss's priority seems evident, but his acrimonious tone and his

---

[17] C. S. Weiss, "On the Methodical and Natural Distribution of the Different Systems of Crystallization," *EPJ*, VIII (1823), 103-110; F. Mohs, "On the Crystallographic Discoveries and Systems of Mohs and Weiss," *ibid.*, pp. 275-290.

implied charge of plagiarism led Mohs to remark that Weiss seemed to intimate that he had "rather invented than abstracted those systems from nature." Mohs, however, had advanced one important step further than Weiss. He recognized the existence of symmetry systems in which the axes were not mutually perpendicular, an idea that led him to establish the present triclinic and monoclinic crystal systems.[18]

Again echoing the ideas of the German *Naturphilosophie*, Mohs asserted that individuality did not require regularity, but implied unity of form. An individual filled the space occupied by that form with a certain matter, thus representing a whole being coherent in itself and limited toward the outside. Minerals upon which the power of crystallization had not exercised its action were without individuality and did not possess any of the properties connected with this state of existence. They lacked unity of space and were not single bodies. Imperfectly formed minerals could be compared to mutilated, defective, or monstrous plants and animals, whereas minerals that were decomposed might be compared to a plant or an animal that has ceased to live. The aim of the science of crystallography, in Mohs's view, was to ascertain the relationships of the regularly limited spaces, that is, the forms of the crystals. Since the concern of crystallography was nothing other than figured space, only geometrical quantities and their relations to one another should be taken into consideration.

Mohs believed that each system of crystallization was based on one fundamental form. These forms were not the primitive forms of Haüy. Mohs's fundamental form was defined as one that could not be derived from another fundamental form, did not possess axes of infinite length, and was contained under the least possible number of faces. The assemblage of simple forms that could be deduced from one fundamental form, independent of all considerations of its dimensions, was a system of crystallization. It was not a mere aggregation of forms of different sorts which had certain geometrical properties peculiar to them. Rather, the system was a grouping of those geometrical relationships that occurred among forms derived from one fundamental form in a completely orderly manner. Interconnecting relations and sequential development of each form in a logical series formed the bases of a crystal

[18] F. Mohs, *Treatise on Mineralogy*, trans. from German by William Haidinger (3 vols.; Edinburgh, 1825).

system. There were six crystal systems, although a seventh was possible, if one divided the hexagonal and rhombohedral divisions of the hexagonal system. Mohs believed, then, that crystal symmetry was founded upon two basic laws of combination. First, the simple derived forms must have a certain relationship with one another which express the peculiarity of the system to which they belong. The symmetry of ensuing combinations of the simple derived forms, then, depends upon the relative dimensions and the position of the simple forms with respect to the axes of the system.

Mohs, more than Weiss, wished to transform crystallography into a pure geometrical science. Both men, however, laid the foundations for the mathematical treatment of the concept of external symmetry which took place during the nineteenth century. It was natural that German scientists assumed the lead in the early stages of this development because of the conceptual background and the work of Weiss and Mohs. Crystal analysis involved the establishment of certain symmetrical groupings of points determined by the rotation of axes. The enclosure of these point groups by plane surfaces gave representations of the possible simple crystal forms. If, in a system where such was allowable, the dimension of one or two axes was doubled, trebled, and so forth, the permissible axial rotations would produce other point groups. The enclosure of these point groups delineated the secondary forms that were possible within a crystal system. The principal task of crystallography was to analyze the operations of symmetry in order to delineate additional classes of the main crystal systems. New point groupings could thus be established to which actual crystalline substances could be referred and classified. In 1830 Johann F. C. Hessel advanced the analysis of external crystal symmetry by demonstrating that, from a geometrical point of view, there can exist only thirty-two point groups, that is, combinations of crystallographic symmetry elements.[19] These are the present crystal classes. Hessel's work appears to have been overlooked, however, for more than half a century,[20] and the thirty-two crystal classes

[19] J. F. C. Hessel, *Krystallometrie oder Krystallonomie und Krystallographie*, Ostwald's Klassiker der exakten Wissenschaften, nos. 88, 89 (Leipzig, 1897). The derivation of the thirty-two crystal classes appears in no. 89, pp. 91-124.

[20] *Ibid.*, no. 88, p. 188. Hessel's contribution was first clearly recognized when L. Sohncke described it in 1891.

were defined independently by Auguste Bravais (1811-1863) in 1848.

With crystallographers primarily interested in the mathematical analysis of symmetry, mineral classification became the work of the mineralogist. He could determine the relative dimensions of the axes of a crystal by measuring the interfacial angles, and, from these dimensions, he could observe whether secondary forms were logically related in accordance with the rules of a particular crystal system. He could refer to the double refracting properties of the system as a further analytical tool in determining the crystal classification of a substance. In addition, chemical analysis was an aid to the mineralogist in the proper identification of the crystal. However, since Mitscherlich had shown that different substances could crystallize in the same form and the same substance could crystallize in different forms, the analysis and determination of the types and quantities of elementary matter present in a crystalline substance were the concern of the chemist who followed the techniques or principles of his particular science. The unified science of crystals which Haüy had visualized, utilizing the results of crystal analysis, chemistry, and mineralogy, was no longer a goal.

The extension of the innovations of Weiss and Mohs also had chauvinistic overtones. In the biographical article on Haüy in the *Biographie universelle* (Michaud), written in the 1850's, Delafosse and Durozier said:

We should, however, recognize that this theory [of Haüy's] has incurred an almost universal disfavor in Germany. The mineralogists of that country, in accepting and in seeking to appropriate the basis of Haüy's ideas, have completely rejected the form. It is necessary to look for the cause of the antipathy that the savants beyond the Rhine have manifested for the molecular theory merely in that idealistic philosophy, that type of nature metaphysics, with which all the intellectuals in Germany are preoccupied. Some revived quibbles of the Greeks which have been refuted a hundred times, some sophisms based on the famous antinomies of Kant, have led the German physicists to prefer, in the study and interpretation of natural phenomena, the kind of vague and obscure explications that they call dynamic rather than the simple, clear, and positive views we deduce from the atomistic hypothesis. They reject every theory in order to limit themselves to experience, or else they put trivial subtleties in place of these representations of phenomena, these constructions of material bodies, which are admitted by the Newtonian philosophy, and which seem to them too mechanistic and too gross because they speak simultaneously to the senses and to reason. In the majority of German works on crystallography, the two fundamental laws of which we have spoken [Haüy's laws of decrement —law of rational intercepts—and Haüy's law of symmetry] are presented as simple empirical laws; they have lost the character of a priori laws, which distinguished them in the theory that Haüy gave us. The law of rational intercepts, for example, has been considered as a consequence of another purely experimental law, which Weiss has called the law of zones. . . . German crystallographers have believed that they could limit themselves to the consideration of the form in crystals, and neglect that of the structure and other physical properties; crystallography in their hands has become again, as it was in the time of Romé de l'Isle, a completely geometric science.[21]

Haüy's followers admitted that the German method had one advantage: it was much easier and more effective in teaching to give a dogmatic exposition of the principles of crystallography to students whose practical experience was limited; the German method was easier to learn. Otherwise they were unsympathetic and opposed to the new method. But the comparison with the work of Romé de l'Isle was hardly fair to the innovations the Germans had made, and the indictment avoided mentioning the progress that had resulted from this fresh approach. Romé de l'Isle's theory of truncations yielded a descriptive system, the first step toward the mathematical analysis of crystals. On the other hand, the systems of Weiss and Mohs had sound foundations in the concept of the presence of symmetry in nature.

The charge was true, however, that the effect of the work of Weiss and Mohs was to concentrate interest on the mathematical relations of external forms. Yet there was always speculation concerning the internal structure of crystals and the shapes of the elements that were the basic constituents. Haüy was not among the scientists who made such postulations. The only reference he made to the physical construction of the integrant molecule was in those instances where two or more substances had such regular shapes as the cube for their integrant molecules. In these instances he proposed that the integrant molecules were constructed of variously shaped elementary molecules whose arrangement chanced to produce the same figure. "Atom" was a word Haüy used rarely, and then only in his late works when referring to Dalton's

---

[21] *Biographie universelle* (Michaud) (52 vols.; Paris, 1853-1866), XVIII, 580.

theory. He termed the theoretical subdivisions of the integrant molecules "elementary molecules," and on no occasion did he assign specific shapes to these particles. It appears that he never attempted to develop the figure of a chemical compound from the figures of the integrant molecules of the elementary substances that composed it. Bernhardi, as an adherent of Weiss's ideas, did do this in an effort to discredit the molecular and advance the polar concept of matter.[22] He pointed out that Haüy had determined the primitive form and the integrant molecule of copper pyrite to be a regular tetrahedron. The elementary chemical constituents were known to be copper, iron, and sulfur. The primitive forms of copper and iron were also known to be regular, that is, either cubic or octahedral, and that of sulfur was an irregular octahedron. In no way, Bernhardi pointed out, could one juxtapose these forms so that a regular tetrahedron would result without postulating empty space between the molecules. The adherents of the molecular theory had no answer for Bernhardi.

Crystals, then, in no way vindicated the molecular concept of matter. They could just as well be thought of as being formed of spherical atoms, and without at least assuming some type of pointlike internal structure, the anisotropy of crystals with respect to their optical and electrical properties was quite inexplicable. The idea of spheres was rejected at about the turn of the nineteenth century on the basis of probability; it was postulated that, other things being equal, the probability that nature would form an elementary molecule in one shape rather than another would vary inversely as the number of planes that terminated them. Hence, between the sphere and the tetrahedron, the probability would be as 4 to infinity.[23]

But despite this appeal to the laws of probability, the hypothesis that the external regularity of crystals might be explained better by the stacking of spherical particles was continually advanced. The writings of Johann Joseph Prechtl (1778-1854) and William Hyde Wollaston illustrate this trend. Both proposed geometrical atomic arrangements within the integrant molecules to account for the regular crystalline structure. Prechtl hypothesized that atoms had no form in the liquid state but took a spherical form

[22] J. Bernhardi, "Gedanken über Krystallogenie, und Anordnung der Mineralien," GJ, VIII (1809), 370.

[23] R. Chenevix, "Réflexions sur quelques méthodes minéralogiques," ADC, LXV (1808), 135.

Fig. 20. Wollaston's illustrations of the packing of spherical particles to account for the structure of crystals. Note the different colored balls in (7), (8), (9), (11), and (14), uniformly intermixed, which Wollaston postulated to account for contemporary views of binary chemical combination.
SOURCE: *W. H. Wollaston, "On the Elementary Particles of Certain Crystals," PT (1813), pl. 2.*

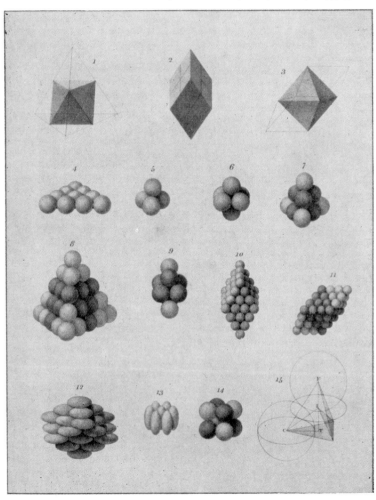

when solidification occurred.[24] The attractive force of each globule emanated from its center of gravity. Equal attractive forces
produced such regular integrant molecules as the tetrahedron or
the octahedron, whereas unequal forces caused the spheres to
become flattened. The grouping of these spheroids resulted in
the formation of irregularly shaped crystals. Wollaston's idea
was independently conceived, but, as he pointed out in a memoir,
he found that the same proposal had been made by Robert Hooke
a century and a half previously.[25] If one supposed that the elementary particles were perfect spheres that by mutual attraction assumed the arrangement that brought them together as closely
as possible, all difficulty in the explanation of the figures of the
integrant molecules would be removed (fig. 20). If one placed
a ball in the cavity produced by the contiguous grouping of three
balls of equal size, a regular tetrahedron would be formed. Similarly an octahedron could be constructed by the suitable arrangement of six contiguous balls. In order to generalize into this concept the theory of binary combination proposed by Dalton and
Berzelius, Wollaston pointed out that four white and four black
balls, the white and the black representing different elements,
might be intermixed so that each would be in equilibrium and
present a cubic configuration. In concluding, Wollaston stated:

And, though the existence of ultimate physical atoms absolutely indivisible may require demonstration, their existence is by no means
necessary to any hypothesis here advanced, which requires merely
mathematical points endued with powers of attraction and repulsion,
equally on all sides, so that their extent is virtually spherical, for from
the union of such particles, the same solids will result as from the combination of spheres impenetrably hard.[26]

Wollaston would not commit himself between the polar and the
molecular theories, and he admitted that attempts to trace any
general correspondence between the internal structure and the
postulated chemical elements of bodies were premature.
   These and many other attempts to construct a model of the
interior structure of crystals from considerations of their cleav-

---

[24] J. Prechtl, "Théorie de la cristallisation," *Journal des Mines*, XXVIII
(1810), 261-312.
[25] W. H. Wollaston, "On the Elementary Particles of Certain Crystals,"
PT (1813), pp. 51-63.
[26] *Ibid.*, p. 61.

age, external symmetry, and directional physical properties were significantly strengthened by a memoir written by Gabriel Delafosse in 1840.[27] It was absolutely necessary, he stated, to revise Haüy's theory by establishing a distinction between the integrant molecule of a crystal and the molecule of its material substance. He continued:

> While the molecule, or better the integrant particle, of a crystal may be considered as representing one of the elements of the mechanical and geometrical structure of the crystal, and not necessarily the atomic element itself, it can be said that its existence is incontestable . . . from the point of view of molecular physics. But its reality is not that which Haüy believed he was able to attribute to it in confusing it with the molecule of the substance; it is important above all to restore this element of the crystal to its true significance.[28]

Admittedly, then, the integrant molecule did not denote the species of a substance, as Haüy had preached. It was rather, Delafosse continued, the representation of the small intermolecular spaces or the outline of the crystal lattice nodes. It was a geometrical element completely distinct from the physical or chemical molecule. It is completely erroneous historically, then, to state that the modern equivalent of Haüy's integrant or subtractive molecule is the unit cell of the space lattice, because such a statement does not take into consideration Haüy's view of the chemistry of the integrant molecule.[29]

Delafosse stressed the idea of an internal crystal lattice which had been postulated previously,[30] and he emphasized that this con-

---

27 G. Delafosse, "Recherches relatives à la cristallisation, considerée sous les rapports physiques et mathématiques," *Académie des Sciences, Comptes Rendus*, XI (1840), 394-400.

28 *Ibid.*, p. 396.

29 G. Delafosse reemphasized this point in his memoir, "Sur la structure des cristaux et ses rapports avec les propriétés physiques et chimiques," *Académie des Sciences, Comptes Rendus*, XLIII (1856), 958-962; and also in *Biographie universelle* (Michaud), XVIII, 579.

30 L. B. Seeber suggested the space lattice in his article, "Versuch einer Erklärung des inneren Baues der festen Körper," *Gilberts Annalen der Physik*, LXXVI (1824), 229-248, 349-372, but, as with Hessel, his ideas seemed to have been overlooked until they were publicized late in the nineteenth century by Quenstedt and Sohncke. Augustin Cauchy also had the notion of the lattice in his article, "Sur les diverses méthodes à l'aide desquelles on peut établir les équations qui representent les lois d'équilibre, ou le movement intérieur des corps solides ou fluides," *Férussac, Bull. Sci. Math.*, XIII (1830), 169-176.

cept would obviate the necessity of pretending there were exceptions to Haüy's law of symmetry. Where such anomalies existed, he said, there must be a different kind of symmetry than had been anticipated. With the divorce of any chemical connotation from Haüy's integrant and substractive molecule and with cleavage viewed as determining only a degree of symmetry and not delineating a specific geometrical figure to which the body must be referred, the theoretical consideration of the internal symmetry of crystals could develop rapidly. Beginning in 1848, Auguste Bravais (1811-1863) wrote a series of papers in which he treated initially the types of geometric figures formed by points distributed regularly in space and then applied these considerations to crystals, with the points viewed as being the centers of gravity of the chemical molecules or as poles of forces.[31] With this approach, Bravais was able to explain the cleavage and external symmetry of crystals as a function of the reticular density. Most important, Bravais demonstrated that there was a maximum of fourteen space lattices or groups of points differing by symmetry and geometry whose translational repetition in space maintained the symmetrical arrangements of the points of a unit cell. He perceived that these fourteen space lattices denoted seven different lattice symmetries which corresponded to the previously recognized seven crystal systems. Hence, the external symmetry became firmly grounded on the concept of the space lattice. Subsequently, the seven systems were reduced to six by considering the rhombohedral system as merely a division of the hexagonal system.

Just how the chemical atoms or molecules were arranged within the unit cells formed by the space lattice, however, remained a matter of speculation. Because it was known that the symmetry of the crystal as a whole was related to the pattern of the internal structure, it was a geometrical problem to determine the number of symmetrical ways of arranging points in space so that the environment around each point was precisely the same as that around any other point, but not necessarily similarly oriented as

---

[31] A. Bravais, "Les systèmes formés par des points distribués regulièrement sur un plan ou dans l'espace," *Journal de l'Ecole Polytechnique*, XIX (1850), 1-128 (presented to the Academy of Sciences, Dec. 11, 1848). There is an English translation of this work by Amos J. Shaler, *On the Systems Formed by Points Regularly Distributed on a Plane or in Space* (Crystallographic Society of America, 1949). Also see A. Bravais, "Etudes cristallographiques," *Journal de l'Ecole Polytechnique*, XX (1851), 101-276 (presented to the Academy of Sciences, Feb. 26, Aug. 6, 1849).

in the space lattice. Leonard Sohncke (1842-1897) was the first
to make progress in this direction.[32] He recognized that the condition of translational equivalence was a restriction justified only
by considering the points from an external orientation. To overcome this difficulty, he introduced two new symmetry elements,
first, the screw axis, in which a rotation around an axis is combined
with a translation of the system along the axis, and second, the
glide plane, in which the reflection in a mirror plane is combined
with a similar translation without rotation along the axis. In this
manner, Sohncke was able to arrive at sixty-five different spatial
arrangements of points.

*Fig. 21. Barlow's predictions of the structures of
simple binary compounds.* SOURCE:
*William Barlow, "A Mechanical Cause of
Homogeneity of Structure and Symmetry,"*
Proceedings of the Royal Dublin Society, *n.s., VIII*
(1897), 547, 549.

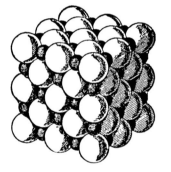

But this result was only a partial answer to the problem. During
the late 1880's, E. S. Federov (1853-1919) arrived at the 230 space
groups, but his work was in Russian and did not become generally
known until after the publication in 1891 of that of Artur Schoenflies, who arrived at the same result by considering the problem as
one of geometrical group theory.[33] The increase in the number of
groups over that attained by Sohncke resulted from the intro-

---

[32] L. Sohncke, *Enwicklung einer Theorie der Kristallstruktur* (Leipzig,
1879).
[33] E. S. Federov, "Simmetriia Pravil'nykh Sistem Figur," *Zap. Min. Obshch.*
["The Symmetry of Real Systems of Configurations," *Transactions of the
Mineralogical Society*], XXVIII (1891), 1-146; A. Schoenflies, *Kristallsysteme
und Kristallstruktur* (Leipzig, 1891).

duction of enantiomorphic considerations. At about the same time, William Barlow (1845-1934), considering the problem from the point of view of a physical model, viewed the atoms in a crystal as spheres that packed themselves into as small a volume as possible.[34] Proceeding from an analysis of the packing of identical atoms, he attempted to visualize the structure that would result from the packing of spheres of different sizes, which might be required for the arrangement of atoms in simple binary chemical compounds (fig. 21). Although this approach was not so systematic as that of Federov and Schoenflies, Barlow did arrive at hypothetical structures of great value to investigators in the early structure determinations by X-ray analysis.

But even with such important advances in mathematical analysis, the actual structure of atoms, ions, or molecules in crystals was agreed to be a matter of speculation. Two crystals belonging to the same crystal class can be constructed from a completely different pattern of internal symmetry elements. It was not possible, then, to correlate an actual crystal with any space group beyond selecting one of those groups which had the same point symmetry as the crystal, and this procedure left many alternatives. The answer to the problem could have been determined, as it finally was, only by the X-ray diffraction technique discovered by Max von Laue.[35]

The focus of interest of crystallographers was completely changed by X-ray diffraction analysis. The primary interest of most crystallographers at the present time does not involve the external crystal form or its symmetry. Instead, the new methods allow investigation of the internal structure of crystalline matter. The atomic and molecular arrangements are studied, which in turn shed light on the relationship between structure and physical properties (figs. 22, 23). It is because of this development that such startling progress in the understanding of matter in the solid state has occurred during the twentieth century.

This is not to say that the science of crystallography is complete, that crystallographers have merely to extend with present

---

[34] William Barlow, "A Mechanical Cause of Homogeneity of Structure and Symmetry," *Proceedings of the Royal Dublin Society*, n.s., VIII (1897), 527-690.

[35] For a complete account of the development of space groups and modern crystallographic techniques, see P. P. Ewald, ed., *Fifty Years of X-ray Diffraction* (Utrecht, 1962).

Fig. 22. Position of the atoms within a unit cell of fluorite, CaF₂, projected onto a cube face. Lettered circles refer to the corresponding spheres in figure 23. Source: Ralph W. G. Wyckoff, Crystal Structures (New York: Interscience Publishers, Inc., 1948), Fig. IV, 1a.

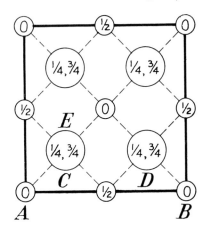

Fig. 23. A perspective packing drawing showing distribution of the atoms of CaF₂ within the unit cube. Atoms have been given their expected ionic sizes. Source: Ralph W. G. Wyckoff, Crystal Structures (New York: Interscience Publishers, Inc., 1948), Fig. IV, 1b.

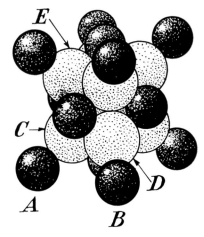

or improved methods their knowledge of the structure and properties of countless chemical substances. It is true that most of the
questions concerning the symmetry and the structure of crystals
posed prior to and during Haüy's time have been sufficiently explained. But the fact, for example, that a number of substances
exist as liquid crystals gives rise to the idea that a higher and
more complete set of generalizations may be found necessary.[36]
Exhibiting double refraction and the ability to polarize light, the
molecules, though mobile, apparently possess a directive force
that arranges them in definite configurations. The knowledge of
the existence of liquid crystals would certainly have intrigued
and delighted Christian Weiss, and they serve to remind us that
the presence of symmetrical forms, not only inorganic but organic,
and the rationale of the processes by which they are formed in
nature yet require much explanation.

---

[36] For consideration of formative processes and symmetry, see D'Arcy
Wentworth Thompson, *On Growth and Form* (Cambridge, 1942); Hermann
Weyl, *Symmetry* (Princeton, 1952); and Lancelot Law Whyte, "A Scientific
View of the 'Creative Energy' of Man," in Morris Philipson, ed., *Aesthetics
Today* (Cleveland, 1961), pp. 349-374.

Appendix I

Haüy's Method of Determining the Relative Dimensions of the Integrant Molecule of Calcite

The following passage is taken from Haüy, *Essai d'une théorie sur la structure des crystaux* (Paris, 1784), pp. 96-97:

"An observation that I made on this same crystal [the hexagonal prism of calcite at the left of fig. 24] furnished me the bases upon which to calculate the plane angles of the nucleus. If, after having detached a segment of the prism by an oblique section, made, for example, in the direction of the plane *amro*, this segment is inverted so that the face *amro* is applied on the part from which it has been detached and so that the line *mr* is identified with the line *ao*, the quadrilateral *zmrq* is found to be even with the plane of the hexagon *dgaonc*, without its being possible to perceive the slightest inclination between the two planes. It follows from this observation that it is extremely likely that the angle resulting from the respective inclination of the two planes *zmrq* and *amro* is exactly equal to the angle that the second of these planes makes with *azqo*; it may therefore be concluded that the triangle *a' c' b'*, formed by the respective inclination of the three planes, is not only right-angled but also isosceles. Now, drawing attention to the fact that the plane *amro* is parallel to the corresponding face of the nucleus, if we suppose that the solid

represented [at the right of fig. 24] is disposed as the nucleus, it
is easy to see that the triangle *a′ c′ b′* is similar to the triangle *ati*,
composed of *ai*, which is half of the small diagonal of the rhombus
acgh, of the line *it* drawn perpendicular to the axis, and of the
portion *at* of the same axis. Then the triangle *ati* is also right-
angled and isosceles.

*Fig. 24. Haüy's calculations of the relative
dimensions of the integrant molecule of calcite.
SOURCE: R. J. Haüy, Essai d'une théorie sur
la structure des crystaux
(Paris, 1784), Pl. II, figs. 16, 21.*

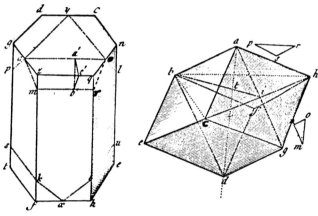

"Let *at*=*it*=1, then *ai*=$\sqrt{2}$; and, since *ti* is the perpendicular
bisector of the equilateral triangle *chb*, formed by the three long
diagonals of the superior faces of the rhombohedron, *ib*=$\sqrt{3}$, and
therefore *ab*= $\sqrt{(ai)^2+(ib)^2}$=$\sqrt{5}$. Solving the right-angled triangle *aib* with the aid of these givens, the logarithm of the sine of
the angle *iab* is .98890756, which corresponds to an angle of 50°
46′ 6″ 30‴. Then *cab*=101° 32′ 13″; and the small angle *abc* is
78° 27′ 47″."

From these values Haüy was able to calculate the value of the
interfacial angles of the rhombohedron to be 104° 28′ 40″ and 75°
31′ 20″. All these calculations were a necessary basis for the re-
mainder of Haüy's theory, the construction of the crystal varieties
by the juxtaposition of the constituent (later integrant and subtractive) molecules and the determination of their facial and
interfacial angles.

Appendix II

Haüy's Construction
of the Octahedron from the Cube

To construct the regular octahedron from the cube as in figure 25, Haüy postulated that it was necessary to apply a law of decrement on the angles A, E, O, I on the upper face and similarly on all angles of the other faces of the cube. He assumed that the cube nucleus was composed of 9 x 9 x 9 or 729 cubic molecules, so that each face showed 81 molecules as in section $EAIO$. As the law commenced to operate, the first two superimposed lamellae took the forms shown in sections $QVGFCLNP$ and $BDHK$. Five integrant molecules were added to the edges, such as EA, forming $QVTS$, and, similarly, additional molecules were added in the next layer at B, D, H, and K. Haüy admitted that this supposition was required because, otherwise, reentrant angles would be formed. The edges of the octahedron, such as sm, would not be straight lines. As molecules were added to the edges of the cube faces, however, they were subtracted from the corners of the cube on lines parallel to the diagonals. The layers reached a maximum length and width of 13 molecules, and, thereafter, each successive superimposed layer decreased by one row of molecules parallel to the diagonals at each of the original angles. The progressively smaller lamellae are shown in figure 25. It is evident that the application of these lamellae would result in the construction of the required octahedron. There were objections that this kind of growth was not postulated in the construction of other forms; in effect, that Haüy was presupposing the secondary surfaces his theory was designed to prove. Haüy's reply was that he did not believe crystalline growth took place exactly in the indicated manner.

Fig. 25. Haüy's construction of the octahedron from the cube. SOURCE: *R. J. Haüy, Traité de minéralogie (5 vols.; Paris, 1801), V, Pl. III, fig. 20; Pl. IV, fig. 23.*

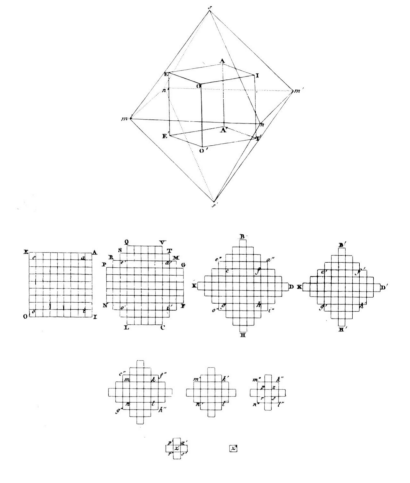

Bibliography

BOOKS

Accum, Frederick. *Elements of Crystallography after the Method of Haüy.* London, 1813.

Achena, M., and Henry Massé. *Avicenne: Le livre de science.* 2 vols. Paris, 1958.

Acton, Alfred, ed. *The Letters and Memorials of Emanuel Swedenborg.* Bryn Athyn, Pa., 1948.

Adam, Charles, and Paul Tannery, eds. *Œuvres de Descartes.* 12 vols. Paris, 1897-1913.

Adams, Frank Dawson. *The Birth and Development of the Geological Sciences.* New York: Dover, 1954.

Baker, Henry. *Employment for the Microscope.* London, 1764.

Barrere, Pierre. *Observations sur l'origine et la formation des pierres figurées.* Paris, 1746.

Bartholin, E. *Versuch mit dem doppeltbrechenden isländischen Kristall.* Leipzig, 1922.

Baumé, Antoine. *Chymie expérimentale et raisonée.* 3 vols. Paris, 1773.

Bergman, Torbern. *Physical and Chemical Essays.* Trans. from Latin by Edmund Cullen. 3 vols. London, 1784.

Bernal, J. D. *Science and Industry in the Nineteenth Century.* London, 1953.

Berthollet, C. L. *Essai de statique chimique.* 2 vols. Paris, 1803.

Bertrand, Elie. *Dictionnaire universel des fossiles propres et des fossiles accidentels.* Avignon, 1763.

Bertrand, Joseph. *L'Académie des sciences et les académiciens de 1666 à 1793.* Paris, 1869.

Berzelius, J. J. *An Attempt To Establish a Pure Scientific System of Mineralogy.* Trans. from Swedish by John Black. London, 1814.

———. *Nouveau système de minéralogie.* Paris, 1819.

Birch, Thomas, ed. *The Works of the Honourable Robert Boyle.* 5 vols. London, 1744.

Boas, Marie. *Robert Boyle and Seventeenth Century Chemistry.* Cambridge, 1958.

Boerhaave, Herman. *A New Method of Chemistry.* London, 1727.

Bonnet, Charles. *Contemplation de la nature.* 2 vols. Geneva, 1770.

Bourguet, Louis. *Lettres philosophiques sur la formation des sels et des cristaux et sur la génération et le méchanisme organique des plantes et des animaux.* Amsterdam, 1729.

Bournon, J. L. de. *Traité complet de la chaux carbonatée et de l'arragonite.* 3 vols. London, 1808.

Boyle, Robert. *The Origine of Formes and Qualities.* 2d ed. Oxford, 1667.

———. *An Essay about the Origine and Virtues of Gems.* London, 1672.

Bravais, M. A. *On the Systems Formed by Points Regularly Distributed on a Plane or in Space.* Trans. from French by Amos J. Shaler. Crystallographic Society of America, 1949.

Brooke, Henry. *A Familiar Introduction to Crystallography: Including an Explanation of the Principle and Use of the Goniometer.* London, 1823.

Browne, Sir Thomas. *Pseudodoxia Epidemica.* London, 1650.

Buffon, George Leclerc, Comte de. *Histoire naturelle des minéraux.* 5 vols. Paris, 1783-1788.

Bulletin de la Société Française de Minéralogie. Vol. LXVII. Paris, 1944.

Caley, Earle R., and John F. C. Richards. *Theophrastus on Stones.* Columbus, Ohio, 1956.

Cappeller, M. A. *Prodromus crystallographiae, de crystallis improprie sic dictis commentarium.* Lucernae, 1723.

———. *Prodromus crystallographiae.* Trans. Karl Mieleitner. München, 1922.

Child, J. M., trans. *R. J. Boscovich: A Theory of Natural Philosophy.* Chicago, 1922.

Childe, V. Gordon. *New Light on the Most Ancient East.* New York, 1957.

Clarke, John, trans. *Rohault's System of Natural Philosophy.* 2 vols. London, 1723.

Cleaveland, Parker. *An Elementary Treatise on Mineralogy and Geology.* Boston, 1822.

Cornford, Francis MacDonald. *Plato's Cosmology.* New York: Library of Liberal Arts, n.d.

Cronstedt, Axel Frederic. *An Essay towards a System of Mineralogy.* Trans. from Swedish by Gustav van Engestrom. 2 vols. London, 1788.

Cuvier, G. *Recueil des éloges historiques.* Paris, 1861.

Cuvier, G., ed. *Rapport Historiques sur le progrès des sciences naturelles depuis 1789 et sur leur état actuel.* Paris, 1810.

Dalton, John. *A New System of Chemical Philosophy.* 2 vols. Manchester, 1808.

Daumas, Maurice, ed. *Histoire de la science.* Bruges, 1957.

Dezallier d'Argenville, A. J. *L'Histoire naturelle: éclaircie dans une de ses parties principales, L'Oryctologie*. Paris, 1755.

Ewald, P. P., ed. *Fifty Years of X-ray Diffraction*. Utrecht, 1962.

France, Anatole, ed. *Les Œuvres de Bernard Palissy*. Paris, 1880.

Freind, John. *Chymical Lectures*. London, 1712.

Gassendi, Petri. *Opera Omnia*. 6 vols. Florence, 1722.

Geoffroy Saint-Hilaire, Isidore. *Vie, travaux, et doctrines scientifiques d'Etienne Geoffroy Saint-Hilaire*. Paris, 1847.

Gerhardt, C. I., ed. *Die philosophischen Schriften von Gottfried Wilhelm Leibniz*. 7 vols. Berlin, 1885.

Gilbert, William. *On the Magnet*. New York, 1958.

Gillispie, C. C. *Genesis and Geology*. Cambridge, Mass., 1951.

Grew, Nehemiah. *Recueil d'expériences et observations sur le combat que procéde du mélange des corps*. Paris, 1679.

———. *The Anatomy of Plants*. London, 1682.

Grignon, P. C. *Mémoires de physique: sur le fer*. Paris, 1775.

Grimaux, Edouard, ed. *Œuvres de Lavoisier*. 6 vols. Paris, 1862-1893.

Groth, Paul. *Entwicklungsgeschichte der mineralogischen Wissenschaften*. Berlin, 1926.

Guglielmini, D. *De salibus dissertatio epistolaris physico-medico-mechanica*. Venetiis, 1705.

Gunther, R. T., ed. *Early Science in Oxford*. 14 vols. Oxford, 1923-1945.

Guyton de Morveau, L. B. *Elémens de chymie: théorique et pratique*. 2 vols. Dijon, 1778.

Halévy, Élie. *England in 1815*. London, 1949.

Harris, John. *Lexicon Technicum*. 2 vols. London, 1710.

Hartmann, Franz. *Paracelsus*. London, 1896.

Hartsoeker, Nicholas. *Recueil des plusiers pièces de physique*. Autrecht, 1722.

Haüy, R. J. *Essai d'une théorie sur la structure des crystaux*. Paris, 1784.

———. *Exposition raisonée de la théorie de l'électricité et du magnetisme*. Paris, 1787.

———. *Traité de minéralogie*. 5 vols. Paris, 1801. 2d ed.: Paris, 1822.

———. *Traité élémentaire de physique*. 2 vols. Paris, 1803.

———. *Tableau comparatif des résultats de cristallographie et de l'analyse chimique relativement à la classification des minéraux*. Paris, 1809.

———. *Traité des caractères physiques des pierres précieuses*. Paris, 1817.

———. *Traité de cristallographie*. 3 vols. Paris, 1822.

Henkel, Johann F. *Pyritologia, or a History of the Pyrites*. London, 1757.

Hessel, J. F. C. *Krystallometrie oder Krystallonomie und Krystallographie*. Ostwald's Klassiker der exakten Wissenschaften, nos. 88, 89. Leipzig, 1897.

Hooke, Robert. *Micrographia*. London, 1665.

Huygens, Christiaan. *Treatise on Light*. Trans. from French by Silvanus P. Thompson. Chicago, 1912.

Jaeger, F. M. *Lectures on the Principles of Symmetry and its Application in All Natural Sciences.* Amsterdam, 1917.

Jowett, B., trans. *The Dialogues of Plato.* 2 vols. New York, 1937.

Kant, Immanuel. *The Metaphysical Foundations of Natural Science.* Trans. from German by Ernest Belfort Bax. London, 1883.

Keill, John. *An Introduction to Natural Philosophy.* London, 1758.

Kepler, Johann. *Strena seu de nive sexangula.* Francofurti ad Moenum, 1611.

Kidd, J. *Outlines of Mineralogy.* 2 vols. Oxford, 1809.

Kirwan, Richard. *Elements of Mineralogy.* London, 1784.

Klug, R. *Des kaiserlichen Mathematikers Johannes Kepler Neujahrsgeschenk oder über die Sechseckform des Schnees.* Linz, 1907.

Kobell, Franx von. *Geschichte der Mineralogie.* München, 1864.

Kuhn, T. S. *The Structure of Scientific Revolutions.* Chicago, 1962.

Lametherie, J. C. de. *Principes de la philosophie naturelle.* Paris, 1805.

Lamy, Bernard. *Traitez de méchanique et de l'équilibre des solides et des liquers.* Paris, 1679.

Larkin, Nathaniel J. *An Introduction to Solid Geometry and to the Study of Crystallography.* London, 1820.

Lasswitz, Kurt. *Geschichte der Atomistik von Mittelalter bis Newton.* Hamburg, 1890.

Laue, Max von. *History of Physics.* Trans. from German by R. Oesper. New York, 1950.

Lee, A. D. P., trans. *Aristotle: Meteorologica.* Cambridge, Mass., 1952.

Lehmann, O. *Molekularphysik, mit besonderer Berüchtsichtigung mikroscopischer Untersuchung und Anleitung zu solchen.* 2 vols. Leipzig, 1888-1889.

Lemery, N. *Cours de chymie.* Paris, 1713.

Leonard, William E., trans. *Lucretius: On the Nature of Things.* New York, 1950.

Lewis, W., ed. *The Chemical Works of Caspar Neumann.* London, 1749.

Linnaeus, Carolus. *Systema naturae.* Holmiae, 1768.

Loemker, Leroy E., trans. and ed. *Gottfried Wilhelm Leibniz: Philosophical Papers and Letters.* 2 vols. Chicago, 1956.

McKeon, Richard, ed. *The Basic Works of Aristotle.* New York, 1941.

Macquer, Pierre J. *Elémens de chymie théorique.* Paris, 1749.

Madden, Edward H. *The Structure of Scientific Thought.* Boston, 1960.

Mairan, J. J. de. *Dissertation sur la glâce.* Paris, 1749.

Mallard, Ernest. *Traité de cristallographie géométrique et physique.* 2 vols. Paris, 1879-1884.

Malus, E. *Théorie de la double réfraction de la lumière dans les substances cristallisées.* Paris, 1810.

Marx, C. M. *Geschichte der Krystallkunde.* Karlsruhe, 1825.

Melsen, A. G. van. *From Atomos to Atom.* New York, 1960.

Merz, John T. *A History of European Thought in the Nineteenth Century.* 4 vols. Edinburgh, 1896-1914.

Metzger, Hélène. *La Genèse de la science des cristaux.* Paris, 1918.

———. *La doctrine chimique en France*. Paris, 1923.

Mitscherlich, A., ed. *Gesammelte Schriften von Eilhard Mitscherlich*. Berlin, 1896.

Mohs, F. *Treatise on Mineralogy*. Trans. from German by William Haidinger. 3 vols. Edinburgh, 1825.

Mornet, D. *Les sciences de la nature en France au XVIIIe siècle*. Paris, 1911.

Musschenbroek, P. *Cours de physique expérimentale et mathématique*. 3 vols. Paris, 1769.

Nagel, Ernest. *The Structure of Science*. New York, 1961.

Naumann, Karl F. *Elemente der theoretischen Krystallographie*. Leipzig, 1856.

Neumann, Franz Ernst. *Beitrage zur Krystallonomie*. Berlin, 1823.

Newton, Sir Isaac. *Opticks*. London, 1730. Reprinted: New York, 1952.

Perrault, C. and P. *Œuvres diverses de physique et de méchanique*. Leyden, 1721.

Quenstedt, Friedrich A. *Grundriss der bestimmenden und technenden Krystallographie*. Tübingen, 1873.

Rackham, H., and W. H. S. Jones, trans. *Pliny: Natural History*. Cambridge, Mass., 1938.

Rendell, James R., and I. Tansley, trans. *The Principia by Emanuel Swedenborg*. 2 vols. London, 1912.

Richard, M. A., ed. *Œuvres complètes de Buffon*. 20 vols. Paris, 1833.

Robinet, J. B. *De la nature*. Amsterdam, 1761.

Romé de l'Isle, J. B. L. *Essai de cristallographie*. Paris, 1772.

———. *L'Action de feu central*. Paris, 1779.

———. *Cristallographie, ou description des formes propres à tous les corps du règne minéral*. 4 vols. Paris, 1783.

Roos, Jacques. *Aspects littéraires du mysticisme philosophique et l'influence de Boehme et de Swedenborg au début du romantisme: William Blake, Novalis, Ballanche*. Strasbourg, 1951.

Rose, Gustav. *Das Krystallo-chemische Mineralsystem*. Leipzig, 1852.

Sage, Balthazar G. *Elémens de minéralogie docimastique*. 2 vols. Paris, 1777.

Sambursky, S. *Physics of the Stoics*. London, 1959.

Schelling, Friedrich von. *Von der Weltseele*. Hamburg, 1798.

Schlieper, Hans. *Emanuel Swedenborgs System der Naturphilosophie*. Berlin, 1901.

Schoenflies, A. *Kristallsysteme und Kristallstruktur*. Leipzig, 1891.

Sigstedt, C. O. *The Swedenborg Epic*. London, 1952.

Singer, Charles, et al., eds. *A History of Technology*. 5 vols.

Smith, Cyril Stanley. *A History of Metallography*. Chicago, 1960.

Sohncke, L. *Entwicklung einer Theorie der Kristallstruktur*. Leipzig, 1891.

Stahl, Georg Ernst. *Traité des sels*. Paris, 1771.

Steno, Nicolaus. *De solido intra solidum naturaliter contento dissertationis prodromus*. Florence, 1669.

———. *Vorläufer einer Dissertation über feste Körper, die innerhalb anderer fester Körper von Natur aus eingeschlossen sind.* Trans. Karl Mieleitner. Ostwald's Klassiker der exakten Wissenschaften. Leipzig, 1923.

Taylor, F. Sherwood. *The Alchemists.* New York, 1949.

Thompson, D'Arcy Wentworth. *On Growth and Form.* Cambridge, 1942.

Thorndike, Lynn. *A History of Magic and Experimental Science.* 8 vols. New York, 1923-1958.

Toksvig, Signi. *Emanuel Swedenborg.* New Haven, 1948.

Toulmin, Stephen. *The Philosophy of Science.* New York, 1960.

Trénard, Louis. *Lyon.* 2 vols. Paris, 1958.

Wallerius, Johann G. *Mineralogie.* Berlin, 1763.

Websky, Martin, et al. *Gedenkworte am Tage der Feier des hundertjährigen Geburtstages von Christian Samuel Weiss den 3 März 1880.* Pamphlet. N.p., [1880?].

Werner, A. G. *On the External Characters of Minerals.* Trans. from German by Albert V. Carozzi. Urbana, Ill., 1962.

Weyl, Hermann. *Symmetry.* Princeton, 1952.

Whittaker, Thomas. *The Neoplatonists.* Cambridge, 1928.

Whyte, Lancelot Law, ed. *Roger Joseph Boscovich.* London, 1961.

Winter, J. G. *The Prodromus of Nicolaus Steno's Dissertation Concerning a Solid Body Enclosed by the Process of Nature within a Solid.* New York, 1916.

ARTICLES   (For journal abbreviations, see list of abbreviations on page viii.)

Aepinus, F. "Mémoire concernant quelques nouvelles expériences électriques remarquables," *Histoire de l'Académie Royale des Sciences et Belles Lettres de Berlin,* XII (1756), 105.

Anonymous. "Eloge de M. Guglielmini," *ADS* (1710), 152-166.

———. "Review of M. Haüy's Traité de Minéralogie," *Edinburgh Review,* III (1803), 42-56.

———. "Discours sur le progrès des sciences, lettres, et arts depuis 1789, jusqu'à ce jour (1808); ou Compte rendu par l'Institut de France à S. M. l'Empereur et Roi," *Edinburgh Review,* XV (1809), 1-24.

———. "Discovery of the Composition of Arragonite," *PM,* XLII (1813), 25-27.

———. "Strontian in Arragonite," *PM,* XLV (1815), 389-391.

———. "Transactions of the Geological Society," *Edinburgh Review,* XXVIII (1817), 174-192.

Arey, M. F. "A Review of the Development of Mineralogy," *Iowa Academy of Science Proceedings,* XIII (1906), 7-14.

Aubisson, M. d'. "Lettre de M. d'Aubisson à M. Berthollet," *ADC,* LXIX (1809), 155-188; LXX (1809), 225-248.

Barlow, William. "A Mechanical Cause of Homogeneity of Structure and Symmetry," *Proceedings of the Royal Dublin Society,* n.s., VIII (1897), 527-690.

Bartholin, E. "Experiments Made on a Crystal-like Body Sent from Iceland," *PT*, V (1670), 2039-2048.

Baumé, A. "Réflexions sur l'attraction et la répulsion qui se manifestent dans la cristallisation des sels," *JDP*, I (1773), 8-10.

Beccaria, John B. "On the Double Refraction of Crystals," *PT*, LII (1761), 486-490.

Benz, Ernst. "Swedenborg als geistiger Wegbahner der deutschen Idealismus und der deutschen Romantik," *Deutsche Vierteljahrschrift für Literaturwissenschaft und Geistesgeschichte*, XIX (1941), 1-32.

Bernhardi, J. "Beobachtung über die doppelte Strahlenbrechung einiger Körper, nebst einigen Gedanken über die allgemeine Theorie derselben," *GJ*, IV (1807), 230-258.

———. "Darstellung einer neuen Methods, Krystallen zu beschreiben," *GJ*, V (1808), 157-198, 492-564, 625-654.

———. "Gedanken über Krystallogenie, und Anordnung der Mineralien," *GJ*, VIII (1809), 360-423.

———. "Beweis dass die Form des Arragonits aus der Grundform des Kalkspaths abgeleitet werden könne," *GJ*, VIII (1809), 152-162.

Berzelius, J. J. "Ueber einige mineralogisch-chemische Gegenstände," *SJ*, XIV (1815), 31-34.

———. "Versuch eines rein chemischen Mineralsystems," *SJ*, XV (1815), 301-363, 419-452.

———. "Anmerkung," *SJ*, XXII (1818), 274-302.

Beudant, F. S. "Recherches tendantes à déterminer l'importance relatives des formes cristallines et de la composition chimique dans la détermination des espèces minérales," *ADC*, IV (1817), 72-84.

———. "Lettre au sujet du mémoire de M. Wollaston," *ADC*, VII (1817), 399-404.

———. "Recherches sur les causes qui peuvent faire varier les formes cristallines d'une même substance minérale," *ADC*, VIII (1818), 5-50.

———. "Lettre de M. Beudant à M. Gay-Lussac sur le mémoire de M. Mitscherlich," *ADC*, XIV (1820), 326-335.

———. "Sur la classification des substances minérales," *ADC*, XXXI (1826), 181-205, 225-243.

Biot, J. B. "Mémoire sur un nouveau genre d'oscillation que les molécules de la lumière éprouvent en traversent certains cristaux," *Mémoires de l'Institut*, I (1812), Part I, pp. 1-371.

———. "Sur le découverte d'une propriété nouvelle dont jouissent les forces polarisent des certains cristaux," *Mémoires de l'Institut*, II (1812), 19-30.

———. "Observations sur la nature des forces qui partagent les rayons lumineux dans les cristaux doués de la double réfraction," *Mémoires de l'Institut* (1813-1815), pp. 221-234.

———. "Mémoire sur l'utilité des lois de la polarisation de la lumière pour reconnaître l'état de cristallisation et de combinaison dans un grand nombre de cas où le système cristallin n'est pas immédiatement observable," *ADS* (1816), pp. 275-346.

Biot, J. B., and L. Thénard. "Mémoire sur l'analyse comparée de l'arragonite et du carbonate de chaux rhomböidal," *GJ*, V (1808), 237-242.

Boas, Marie. "The Establishment of the Mechanical Philosophy," *Osiris*, X (1952), 412-541.

Borlase, William. "An Enquiry into the Original State and Properties of Spar and Sparry Productions, Particularly the Spars of Crystals Found in the Cornish Mines, Called Cornish Diamonds," *PT*, XLVI (1749), 250-277.

Bournon, J. L. de. "Sur quelques points de cristallographie," *JDP*, LXXI (1810), 364-366.

Bravais, A. "Les systèmes formés par des pointes distribués regulièrement sur un plan ou dans l'espace," *Journal de l'Ecole Polytechnique*, XIX (1850), 1-128.

———. "Etudes cristallographiques," *Journal de l'Ecole Polytechnique*, XX (1851), 101-276.

Brewster, David. "On the Affections of Light Transmitted through Crystallized Bodies," *PT* (1814), Part I, pp. 187-218.

———. "On the Laws of Polarization and Double Refraction in Regularly Crystallized Bodies," *PT* (1818), pp. 199-273.

———. "On the Difference between the Optical Properties of Arragonite and Calcareous Spar," *Quarterly Journal of Science*, IV (1818), 112-114.

———. "Additions aux observations sur les rapports entre la forme primitive des minéraux et le nombre de leur axes de double réfraction," *JDP*, XCI (1820), 300-309.

———. "On the Connection between the Optical Structure and the Chemical Composition of Minerals," *EPJ*, V (1821), 1-8.

———. "On the Form of the Integrant Molecule of Carbonate of Lime," *Geological Society Transactions*, V (1821), 83-86.

———. "Reply to Mr. Brooke's Observations on the Connexion between the Optical Structure of Minerals and Their Primitive Forms," *EPJ*, IX (1823), 361-372.

———. "Observations on the Pyro-Electricity of Minerals," *Edinburgh Journal of Science*, I (1824), 208-215.

———. "Observations Relative to the Motions of Molecules of Bodies," *Edinburgh Journal of Science*, X (1829), 215-222.

———. "On the Production of Regular Double Refraction in the Molecules of Bodies by Simple Pressure; with Observations on the Origin of the Doubly Refracting Structure," *PT* (1830), pp. 87-95.

Brooke, H. J. "Observations on a Memoir by the Abbé Haüy on the Measurement of the Angles of Crystals," *AP*, XIV (1819), 453-456.

Buée, A. Q. "Outlines of the Mineralogical Systems of Romé de l'Isle and the Abbé Haüy: with Observations," *Nicholson's Journal*, IX (1804), 26-39, 78-88.

Burmester, L. "Geschichtliche Entwicklung des kristallographischen Zeichens und dessen Ausführung in schrager Projection," *Zeitschrift für Kristallographie*, LVII (1922), 1-47.

Canton, John. "Letter to Benjamin Franklin," *PT*, LIII (1762), 457.

Carangeot, A. "Goniomètre, ou mésure-angle," *JDP*, XXII (1783), 193.

———. "Lettre à M. de Lametherie sur le goniomètre," *JDP*, XXIX (1786), 226-227.

———. "Lettre de M. Carangeot à M. Kaestner," *JDP*, XXXI (1787), 204.

Cauchy, A. "Sur les diverses méthodes à l'aide desquelles on peut établir les équations qui representent les lois d'équilibre, ou le movement intérieur des corps solides ou fluides," *Férussac, Bull. Sci. Math.*, XIII (1830), 169-176.

Chenevix, R. "Remarques sur un ouvrage intitulé Materialen zu einer chemie des neunzehnten Jahrhunderts etc. ou materiaux pour servir de base à une chimie du dix-neuvième siecle, publiée par D. J. B. Oersted, Ratisbonne, 1803," *ADC*, L (1804), 173-199.

———. "Observations sur une mémoire du docteur Christian Samuel Weiss, imprimé dans la Minéralogie de M. Haüy, traduite en allemand par Dietrich-Ludwig-Gustave Karsten, conseiller des mines de S. M. Prussienne, etc., avec quelques remarques sur cette traduction," *ADC*, LII (1804), 308-339.

———. "Réflexions sur quelques méthodes minéralogiques," *ADC*, LXV (1808), 5-43, 113-160, 225-277.

Cuvier, G. "Analyse des travaux de l'Académie royale des sciences pendant l'année 1817, partie physique," *ADS* (1817), pp. xciii-clxix.

———. "Analyse des travaux de l'Académie royale des sciences pendant l'année 1818, partie physique," *ADS* (1818), pp. clxxix-ccxxx.

Daubenton, L., and P. Laplace. "Rapport de l'Académie des sciences sur l'essai d'une théorie sur la structure des crystaux," *JDP*, XXIV (1784), 71-74.

Delafosse, G. "Recherches relatives à la cristallisation considerée sous les rapports physiques et mathématiques," *Académie des Sciences, Comptes Rendus*, XI (1840), 394-400.

———. "Sur la structure des cristaux et ses rapports avec les propriétés physiques et chimiques," *Académie des Sciences, Comptes Rendus*, XLIII (1856), 958-962.

Dingle, Herbert. "The Scientific Work of Emanuel Swedenborg," *Endeavor*, XVII (1958), 127-132.

Donnay, J. D. H., and David Harker. "A New Law of Crystal Morphology Extending the Law of Bravais," *American Mineralogist*, XXII (1937), 446-467.

Federov, E. S. "Simmetriia Pravil'nykh Sistem Figur," *Zap. Min. Obshch.* ["The Symmetry of Real Systems of Configurations," *Transactions of the Mineralogical Society*], XXVIII (1891), 1-146.

Fischer, Emil. "Christian Samuel Weiss und seine Bedeutung für die Entwicklung der Krystallographie," *Wissenschaftliche Zeitschrift der Humboldt-Universität zu Berlin*, XI (1962), 249-255.

———. "Christian Samuel Weiss und die zeitgenössische Philosophie (Fichte, Schelling)," *Forschungen und Fortschritte*, XXXVII (1963), 141-143.

Friedrich, W., P. Knipping, and M. von Laue. "Interferenzen-Erscheinungen bei Röntgenstrahlen," *Sitzungsberichte der mathematisch-physikalischen Klasse der K. B. Akademie der Wissenschaften zu München* (1912), pp. 303-322.

Fuchs, J. "Ueber den Gehlenit, ein neues Mineral aus Tirol," *SJ*, XV (1815), 377-386.

————. "Ueber einiger phosphorsauere Verbindungen," *SJ*, XVIII (1816), 288-296.

————. "Ueber den Arragonit und Strontianit," *SJ*, XIX (1817), 113-137.

Gay-Lussac, J. L. "De l'influence de la pression de l'air sur la cristallization des sels," *ADC*, LXXXVII (1813), 225-236.

Geoffroy, M. "Sur l'orgine des pierres," *ADS* (1716), p. 8.

Gillispie, C. C. "The Discovery of the Leblanc Process," *Isis*, XLVIII (1957), 152-170.

Grew, Nehemiah. "On the Nature of Snow," *PT*, VIII (1673), 5193-5196.

Hall, Sir James. "Account of a Series of Experiments Shewing the Effects of Compression in Modifying the Action of Heat," *Philosophical Transactions of the Royal Society of Edinburgh*, VI (1812), 71-183.

Haüy, R. J. "Extrait d'une mémoire sur la structure des cristaux de grenat, presenté a l'Académie royale des sciences et approuvée par cette compagnie le 21 fevrier 1781," *JDP*, XIX (1782), 366-370.

————. "Extrait d'une mémoire sur la structure des spaths calcaires, approuvée par l'Académie royale des sciences le 22 décembre 1781," *JDP*, XX (1782), 33-39.

————. "Mémoire sur la structure des cristaux de feld-spath," *ADS* (1784), pp. 273-286.

————. "Mémoire sur les propriétés électriques de plusiers minéraux," *ADS* (1785), pp. 206-209.

————. "Mémoire sur la structure des divers cristaux métalliques," *ADS* (1785), pp. 213-228.

————. "Lettre de M. l'Abbé Haüy à M. de Lametherie sur le schorl blanc," *JDP*, XXVIII (1786), 63-64.

————. "Mémoire sur la structure du cristal de roche," *ADS* (1786), pp. 78-94.

————. "Mémoire où l'on expose une méthode analytique pour résoudre les problèmes relatifs à la structure des cristaux," *ADS* (1788), pp. 13-33.

————. "Mémoire sur la double réfraction du spath d'Islande," *ADS* (1788), pp. 34-61.

————. "Mémoire sur la manière de ramener à la théorie du parallélipipède celle de toutes les autres formes primitives des cristaux," *ADS* (1789), pp. 519-532.

————. "Exposition abregée de la théorie de la structure des cristaux," *ADC*, III (1789), 1-28.

————. "Traité des caractères extérieurs des fossiles, traduit de l'allemand de M. A. G. Werner, par le tradacteur des Mémoires de chimie de Scheele, 1790," *ADC*, IX (1791), 174-192.

————. "Sur la double réfraction du spath calcaire transparent," *JHN*, I (1792), 63; II (1792), 158-160.

————. "Sur le diamant," *JHN*, I (1792), 377-384.

————. "Sur la double réfraction du cristal de roche," *JHN*, I (1792), 406-408.

———. "Extrait des observations sur la vertu électrique que plusiers minéraux acquièrent à l'aide de la chaleur," *JHN*, I (1792), 449-461.

———. "De la structure considerée comme caractère distinctif des minéraux," *JHN*, II (1792), 56-71.

———. "Sur la double réfraction de plusiers substances minérales," *ADC*, XVII (1793), 140-156.

———. "Exposition de la théorie sur la structure des cristaux," *ADC*, XVII (1793), 225-319.

———. "Mémoire sur les méthodes minéralogiques," *ADC*, XVIII (1793), 225-240.

———. "Mémoire sur de nouvelle variétés de chaux carbonatée avec quelques observations sur les erreurs auxquelles on s'expose en se bornant à l'usage du goniomètre, pour la description des cristaux," *AMHN*, I (1802), 114-126.

———. "Sur l'arragonite," *AMHN*, XI (1808), 241-270.

———. "Addition au mémoire sur l'arragonite," *AMHN*, XIII (1809), 241-253.

———. "Observations sur la simplicité des lois auxquelles est soumise la structure des cristaux," *AMHN*, XVIII (1811), 169-205.

———. Mémoire sur une loi de cristallisation, appelée *loi de symétrie*," *Mémoires du Museum d'Histoire Naturelle*, I (1815), 81-101, 206-225, 273-298, 341-352.

———. "Sur l'électricité produite dans les minéraux," *Mémoires du Museum d'Histoire Naturelle*, III (1817), 223-228.

———. "Comparaison des formes cristallines de la strontiane carbonatée avec celles de l'arragonite," *ADC*, V (1817), 439-441.

———. "Observations sur la mesure des angles des cristaux," *JDP*, LXXXVII (1818), 233-253.

Hill, John. "Spatogenèsie, ou traité de la nature et de la formation du spath," *JDP*, III (1774), 209-218.

Hire, Gabriel de la. "Observations sur une espèce de talc qu'on trouve communément proche de Paris au-dessus des bancs de pierre de plâtre," *ADS* (1710), pp. 341-352.

Homberg, W. "Essai de chimie," *ADS* (1702), pp. 33-52.

———. "Mémoire touchant les acides," *ADS* (1708), pp. 312-323.

Hooykas, R. "La cristallographie dans l'Encyclopédie," *Revue d'Histoire des Sciences*, IV (1951), 344-351.

———. "Torbern Bergman's Crystal Theory," *Lychnos* (1952), pp. 21-54.

———. "The Species Concept in Eighteenth-Century Mineralogy," *Archives Internationales d'Histoire des Sciences*, XXXI (1952), 45-55.

———. "La naissance de la cristallographie en France au XVIIIᵉ siècle," *Les Conferences du Palais de la Découverte*, Ser. D, no. 21 (1955).

———."Les débuts de la théorie cristallographique de R. J. Haüy d'après les documents originaux," *Revue d'Histoire des Sciences*, VIII (1955), 319-337.

Irving, J. A. "Leibniz' Theory of Matter," *Philosophy of Science*, III (1936), 208-214.

Jussieu, A. de. "Réflexions sur plusiers observations concernant la nature de gypse," *ADS* (1719), 82-93.

Keir, James. "On the Crystallizations Observed on Glass," *PT*, LXVI (1776), 530-542.

Kraus, Edward H. "Haüy's Contribution to Our Knowledge of Isomorphism," *American Mineralogist*, III (1918), 126-130.

Kunz, George F. "The Life and Work of Haüy." *American Mineralogist*, III (1918), 61-89.

Lacroix, Alfred. "La vie et l'œuvre de l'abbé René Just Haüy," *Bulletin de la Société Française de Minéralogie*, LXVII (1944), 15-94.

Lametherie, J. C. de. "Mémoire sur la cristallisation," *JDP*, XVII (1781), 251-265.

———. "Sur la vie et ouvrages de M. Romé de l'Isle," *JDP*, XXXVI (1790), 315-323.

Lana, Francis. "On the Formation of Crystals," *PT*, VII (1672), 4068.

Laue, M. von. "Historical Introduction," in *International Tables for X-ray Crystallography*. Vol. I, pp. 1-5. Birmingham, 1952.

Laugier, A. "Note sur la présence de la strontiane dans l'arragonite," *Mémoires du Museum d'Histoire Naturelle*, I (1815), 66.

Leblanc, N. "Essai sur quelques phénomènes relatifs à la cristallisation des sels neutres," *JDP*, XXVIII (1786), 341-345.

———. "Observations générales sur les phénomènes de la cristallisation," *JDP*, XXXIII (1788), 374-379.

———. "De la Cristallotechnie ou Essai sur les phénomènes de la cristallisation," *JDP*, LV (1802), 296-313.

Leeuwenhoek, Anthony van. "Concerning the Various Figures of the Salts Contained in the Several Substances," *PT*, XV (1685), 1073.

———. "On the Dissolution of Silver," *PT*, XXIII (1703), 1430-1443.

———. "Concerning the Figures of Sand," *PT*, XXIV (1704), 1537-1555.

———. "Concerning the Figures of the Salts of Crystals," *PT*, XXIV (1705), 1906-1917.

———. "On the Particles of Crystallized Sugar," *PT*, XXVI (1709), 479-484.

———. "Microscopical Observations upon the Configuration of Diamonds," *PT*, XXVI (1709), 479-484.

———. "Microscopical Observations on the Crystallized Particles of Silver Dissolved in Aqua Fortis," *PT*, XXVII (1710), 20-23.

Lemery, M. "Explication méchanique de quelques différences assez curieuses qui résultent de la dissolution de différens sels dans l'eau commune," *ADS* (1716), pp. 154-172.

M***, M. "Histoire des opinions philosophiques sur les principes et les élémens des corps," *JDP*, X (1777), 286-300.

Malus, E. "Sur une propriété de la lumière réfléchie," *Mémoires de Physique et de Chimie de la Société d'Arcueil*, II (1809), 143-158.

———. "Sur une propriété des forces répulsives qui agissent sur la lumière," *Mémoires de Physique et de Chimie de la Société d'Arcueil*, II (1809), 254-267.

Millington, E. C. "Theories of Cohesion in the Seventeenth Century," *Annals of Science*, V (1941-1947), 253-269.

———. "Studies in Capillarity and Cohesion in the Eighteenth Century," *Annals of Science*, V (1941-1947), 352-369.

Mitscherlich, E. "Ueber die Kristallisation der Salze in denen des Metall der Basis mit zwei proportionen Sauerstoff verbunden ist," *ADB* (1818-1819), pp. 427-437.

———. "Ueber das Verhältnis der Kristallform zu den chemischen Proportionen," *ADB* (1822-1823), pp. 25-41.

———. "Ueber die Körper, welche in zwei verschiedenen Formen kristallisieren," *ADB* (1822-1823), pp. 43-48.

Mohs, F. "On the Crystallographic Discoveries and Systems of Mohs and Weiss," *EPJ*, VIII (1823), 275-290.

———. "General Reflections on Various Important Subjects in Mineralogy," *EPJ*, XIII (1825), 205-218.

Parsons, J. "An Account of Perfect Minute Crystal Stones," *PT*, XLIII (1745), 468-472.

Partington, J. R. "Origins of the Atomic Theory," *Annals of Science*, IV (1939), 245-282.

Pelletier, M. "Observations sur la cristallisation artificielle du soufre et du cinabre," *JDP*, XIX (1782), 311.

Petit, M. "Mémoire sur la végétation des sels," *ADS* (1722), pp. 95-116.

Prechtl, J. "Théorie de la cristallisation," *Journal des Mines*, XXVIII (1810), 261-312.

Réaumur, R. A. de. " Sur la nature et la formation des cailloux," *ADS* (1721), pp. 255-276.

———. "Sur le rondeur que semblent affecter certaines espèces de pierre et entre autres sur celle qu'affecte le caillou," *ADS* (1723), pp. 273-284.

———. "De l'arrangement que prennent les parties des matières métalliques et minérales, lorsqu'après avoir été mises en fusion, elles viennent à se figer," *ADS* (1724), pp. 307-316.

Rouelle, G. F. "Mémoire sur les sels neutres, dans lequel on propose une division méthodique de ces sels, qui facilite les moyens pour parvenir à la théorie de leur cristallisation," *ADS* (1744), pp. 352-364.

———. "Sur le sel marin," *ADS* (1745), pp. 57-79.

Schneer, Cecil. "Kepler's New Year's Gift of a Snowflake," *Isis*, LI (1960), 531-545.

Seeber, L. B. "Versuch einer Erklärung des inneres Baues der festen Körper," *Gilberts Annalen der Physik*, LXXVI (1824), 229-248, 349-372.

Thomson, Thomas. "Some Observations in Answer to Mr. Chenevix's Attack upon Werner's Mineralogical Method," *AP*, I (1813), 241-258.

Tournefort, J. P. de. "Description du labyrinthe de Candie, avec quelques observations sur l'accroissement et la génération des pierres," *ADS* (1702), pp. 217-234.

Vauquelin, L. "Pour déterminer les rapports de l'acide carbonique dans les carbonates de chaux, de baryte, de strontiane, dans l'arragonite, le cuivre bleu et le cuivre vert de Chessy, suivis de l'analyse de l'arragonite d'Auvergne," *ADC*, XCII (1814), 311-319.

Vauquelin, L., and A. Fourcroy. "Expériences comparées sur l'arragonite d'Auvergne et le carbonate de chaux d'Islande," *AMHN*, IV (1804), 405-411.

Weiss, C. S. "Dissertatio de Indagando formarum cristallinarum caractère geometrico principali, ou Mémoire sur la détermination du caractère geométrique principal des formes cristallines," *Journal des Mines*, XXIX (1811), 349-391, 401-444.

———. "Uebersichtliche Darstellung der verschiedenen natürlichen Abteilungen der Kristallisations-systeme," *ADB* (1814-1815), pp. 289-344.

———. "Ueber ein verbesserte Methode für die Bezeichnung der verschiedenen Flächen eines Kristallisations-systeme, nebst Bemerkungen über den Zustand von Polarisirung der Seiten in den Linien der kristallinischen Struktur," *ADB* (1816-1817), pp. 286-314.

———. "Betrachtung der Dimensionsverhältnisse in den Hauptkörpern des sphäröedrischen Systems and ihren Gegenkörpern, in Vergleich mit den harmonischen Verhältnissen der Töne," *ADB* (1818-1819), pp. 227-241.

———. "On the Methodical and Natural Distribution of the Different Systems of Crystallization," *EPJ*, VIII (1823), 103-110.

Whitlock, Herbert P. "René Just Haüy and His Influence," *American Mineralogist*, III (1918), 92-98.

Whyte, Lancelot Law. "A Scientific View of the 'Creative Energy' of Man." In Morris Philipson, ed., *Aesthetics Today*. Cleveland, 1961.

Wilson, Benjamin. "Experiments on the Tourmalin," *PT*, LI (1759), 308-313.

———. "Observations upon Some Gems Similar to the Tourmalin," *PT*, LII (1761), 443-447.

Winterl, J. J. "Critik der Hypothese welche das gegenwärtige Zeitalter der Naturwissenschaft (Physik, Chemie, und Physiologie) zum Grunde legt," *GJ*, VI (1808), 1-35, 201-270.

Wollaston, W. H. "On the Oblique Refraction of Iceland Crystal," *PT* (1802), pp. 381-386.

———. "The Description of a Reflective Goniometer," *PT* (1809), pp. 253-258.

———. "On the Primitive Crystals of Carbonate of Lime, Bitter Spar, and Iron Spar," *PT* (1812), pp. 159-162.

———. "On the Elementary Particles of Certain Crystals," *PT* (1813) pp. 51-63.

———. "Observations on M. Beudant's Memoir 'Sur la détermination des espèces minérales,'" *AP*, XI (1818), 283-286.

Index